P9-DEY-737

MOLLUSC DISEASES

GUIDE FOR THE SHELLFISH FARMER

WSG-I5 90-03

MOLLUSC DISEASES

GUIDE FOR THE SHELLFISH FARMER

Ralph A. Elston

Washington Sea Grant Program
Distributed by University of Washington Press
Seattle and London

This guide to mollusc diseases is the result of cooperation among several institutions. Battelle Marine Sciences Laboratory, Sequim, Washington, provided the support needed to write the guide. Additional support was provided by the U.S. Department of Energy under Contract DE-AC06-76RLO 1830 to Pacific Northwest Laboratories. Editorial and design work was supported by NOAA Grant NA899AA-D-SG022 to the Washington Sea Grant Program, project A/PC-5. The Washington State Department of Fisheries funded publication of the guide under an appropriation for shellfish studies from the Washington state legislature.

The State of Washington and the U.S. Government are authorized to produce and distribute reprints for governmental purposes notwithstanding any copyright notation that may appear hereon.

All rights reserved. No part of this publication may be reproduced or transmitted in any form or by any means, electronic or mechanical, including photocopying, recording, or any information storage or retrieval system, without permission in writing from the author.

Figures of hard-shell clam and sea scallop are redrawn from R. D. Barnes, *Invertebrate Zoology,* 5th ed. (1987), published by Saunders College Publishing, Philadelphia, and used with their permission.

© 1990 University of Washington. Printed in the United States of America.

Cover photo: Oyster larva with OVVD (oyster velar virus disease).

Library of Congress Cataloging-in-Publication Data
Elston, R. A.
      Mollusc diseases : guide for the shellfish farmer / Ralph A.
  Elston
      p. cm.
      Includes bibliographic references.
      ISBN 0-295-97001-4 : $9.95
      1. Mollusks—Diseases. 2. Bivalvia—Diseases. I. Washington Sea
Grant Program. II. Title.
SH179.S5E44 1990
639'.4—dc20
                                  90-12053
                                  CIP

ACKNOWLEDGMENTS

I am grateful to the many individuals who encouraged and helped me with this work. David Alderman, Jim Donaldson, Susan Ford, Herb Hidu, Michael Kent, Ted Meyers, John Pitts, Albert K. Sparks, and Dick Wilson carefully reviewed the entire work; Susan Bower and Gene Burreson kindly reviewed parts of the manuscript. Paul Van Banning provided technical papers and translations on shell disease. Ann Trelstad carefully prepared the typescript. Marilyn Wilkinson provided careful and repeated editorial review of several drafts.

The publication of a work such as this was enthusiastically encouraged by Judith Freeman and Dick Burge of the Washington Department of Fisheries and John Pitts of the Washington Department of Agriculture. Special thanks are due to Ken Chew of the University of Washington for his interest and support. The attention given to publication by the Washington Sea Grant staff is appreciated, especially the careful editing of the manuscript by Alma Johnson and development of the illustrations and cover by Vicki Loe.

I also offer what can only be token acknowledgment for the encouragement given by my wife Heidi for my professional activities, in the face of her own full-time professional and family commitments.

CONTENTS

About the Guide vii

Notable Oyster Diseases 1
"Dermo" (Perkinsiosis) of the American (Eastern) Oyster *(Crassostrea virginica)* **1**
MSX Disease of the American Oyster **4**
Seaside Haplosporidiosis of the American Oyster **8**
Velar Virus Disease of the Pacific Oyster *(Crassostrea gigas)* **9**
Denman Island Disease of the Pacific Oyster **12**
Nocardiosis of the Pacific Oyster **14**
Bonamiasis of the European Flat Oyster *(Ostrea edulis)* **17**
Marteiliasis of the European Flat Oyster **21**
Gill Disease of the Portuguese Oyster *(Crassostrea angulata)* **22**
Hexamitiasis of *Ostrea* and *Crassostrea* Oysters **24**
Shell Disease of Oysters **25**

Other Diseases, Other Molluscs 29
Hemic Neoplasia of Bivalve Molluscs **29**
Vibriosis of Larval and Juvenile Molluscs **32**
Hinge Ligament Disease of Juvenile Bivalve Molluscs **35**
Ameboflagellate Disease of Larval Geoduck Clams *(Panope abrupta)* **37**
Diseases of Abalone **39**

Less Documented Diseases 41
Rickettsia and Chlamydia of Molluscs **41**
Nuclear Inclusion X (NIX) of Pacific Razor Clams *(Siliqua patula)* **42**
Malpeque Bay Disease of American Oysters **43**
Gill Parasite of the Japanese Scallop *(Patinopectin yessoensis)* **43**
Miscellaneous Diseases **45**

Anatomy of Bivalve Molluscs 49

Preventing and Managing Disease in the Hatchery 57
Bacteriological Sampling **58**

Seeking Professional Assistance 63
Chemical Preservation of Tissues **63**
Shellfish Pathology Services **64**

Glossary 71

ABOUT THE GUIDE

The preparation of this guide to the important known infectious diseases of molluscs of commercial importance is the result of many requests from shellfish growers for information on the risks, distribution, prevention and management of diseases. As husbandry for any species or type of animal develops, the significant role of infectious diseases in decreasing productivity and product quality is increasingly recognized. Numerous examples worldwide demonstrate that entire shellfish industries in large coastal regions can be eliminated as the result of shellfish diseases.

The purpose of this guide is to enable shellfish farmers to educate themselves with regard to important diseases of the molluscs they culture and to develop an approach for the control of these diseases. Often, shellfish and fish farmers speak of "natural mortality." Many times this natural mortality may reach a level of 50% of the standing crop of animals within one year. The concept of natural mortality is really nothing more than the acceptance of deaths of animals as a phenomenon over which the farmer has no control. Every death of an individual farmed animal has some biological explanation, although we are not always perceptive enough to discern the cause. So, in fact, there is no such thing as natural mortality. Large-scale deaths of farmed animals are often due to infectious diseases—that is, due to diseases caused by microorganisms such as viruses, bacteria, fungi, or parasites. Many of these deaths can be prevented or managed. It is to this concept that this guide is dedicated.

By familiarizing themselves with the concept of infectious diseases and the way such diseases can be spread by poor practices and prevented or managed by good practices, shellfish farmers can improve the productivity and profitability of their operations. In some cases, this requires that the farmer take the long-term view and sacrifice short-term gains. For example, in some cases, it is wiser to farm indigenous strains of shellfish than to risk introduction of infectious diseases by importing exotic shellfish. On the other hand, it is the responsibility of those of us practicing shellfish pathology to find solutions for problems posed by infectious diseases.

Along these lines, it is the philosophy of this guide that moving shellfish from one geographic area to another, often necessary in their commerce, can frequently be done with little risk of spreading infectious disease if certain precautions are taken. The enforcement of precautionary measures is usually the role of state, provincial, or federal government. However, government and industry should have a similar, if not an identical, objective with respect to infectious diseases of molluscs: the preservation and productive use of shellfish, free of the potentially devastating effects of disease.

This objective can be met only if government and industry recognize that they share a common goal. It is clear that government regulations regarding the control of shellfish diseases are essentially unenforceable and useless unless the industry supports them. Thus, it is the responsibility of government to develop workable policies and effective means of implementation; it is the responsibility of individuals in industry to understand the potential consequences of infectious diseases and to promote this recognition throughout the industry.

Organization of the Guide

This guide does not mention all of the known infectious diseases of molluscs. It does provide a summary of the major facets of the most important diseases. The emphasis is on bivalves, the primary group in commercial cultivation today. The guide is organized by species and by disease. Each treatment of a major disease includes an historical summary, information on its geographic distribution, which species it infects, mortality rate, environmental factors, seasonality, diagnosis, and, most important, prevention and management. Not all of this information is available for each disease, because the science of health management and disease control of bivalve molluscs is in a relatively primitive state today. As the industry develops, the science and knowledge base for health management will also increase.

Because so little is known about some diseases, they are treated in an abbreviated form. Abalones, for example, are increasingly important but little is known about the diseases of these animals. Some diseases are important only from historical interest or because of their impact on an unfarmed natural population of molluscs. Short summaries of some of these diseases are included for general background information in the "miscellaneous diseases" section. Technical references are given after each section. The literature can be retrieved from most university libraries if it is needed for further reference.

I have omitted from the guide many diseases that are mentioned only briefly in the technical literature, particularly those that affect wild, unfarmed species. Since so little is known about these, including them would complicate the simplicity that I believe is necessary to make this guide useful.

There is one last point on this subject. Because no diseases are reported for a particular species does *not* mean that the species has no important diseases. It probably means that the species has only recently been farmed and that its diseases have not been studied. Although many diseases exist in wild populations of molluscs, they are often not recognized until someone attempts to farm the mollusc.

The guide contains several figures on the anatomy of bivalve molluscs. A knowledge of the anatomy is invaluable in understanding biology and disease processes.

Hatchery managers can do much to prevent disease or mitigate its effects in hatchery operations. In addition to the sections on prevention and management specific to discrete diseases, the guide offers general guidelines for preventing and managing disease and detailed instructions for bacteriologic sampling throughout the hatchery system to test for the presence and abundance of bacteria.

Accurate diagnosis of shellfish diseases often depends on the services of a pathologist. The section on professional assistance describes the way to prepare, preserve, and present diseased animals or animal tissues for laboratory investigation and lists the pathology services now known to be available.

A glossary at the end of the guide defines some of the terms common to disease management and pathology. Pathology, like any discipline, uses many terms that can intimidate the uninitiated but, in reality, represent easily understood concepts.

The references at the end of each section are the scientific foundation from which each section was written. Thus, each discussion of a particular disease represents many years of research effort by the scientists listed in the references.

NOTABLE OYSTER DISEASES

"Dermo" (Perkinsiosis) of the American (Eastern) Oyster *(Crassostrea virginica)*

"Dermo" (or, more formally, perkinsiosis) is caused by a parasite (now named *Perkinsus marinus,* but previously known both as *Dermocystidium marinum* and as *Labyrinthomyxa marina)* that infects almost all tissues of the oyster. It is transmitted by direct contact in water but is limited by its inability to tolerate low salinities and low temperatures.

The disease occurs during the warmest months of the year and is more severe in highly concentrated populations of oysters. The disease has had a serious impact on oyster culture on the Gulf and Atlantic coasts for at least 45 years, first being described in association with serious mortalities in the Gulf of Mexico in the 1940s but probably present as early as the 1930s.

Geographic Range and Species Infected

The disease occurs on the Atlantic coast of the United States from Delaware Bay south and on the Gulf Coast. However, it does not occur in all oyster-growing areas within this general geographic range. It has been reported in American oysters cultured in Hawaii, but it is not known if the disease is still present there. The severe effects of this disease in Chesapeake Bay and along most of the Gulf Coast are especially well known. There is a resurgence of the disease in Chesapeake Bay associated with drought.

The American oyster is the only species known to be infected by this parasite, although there are reports of similar parasites infecting other molluscs elsewhere in the world.

Mortality Rate, Environmental Factors, and Seasonality

The severity of the disease increases periodically in infected areas. Mortalities can reach 100% and have been reported to be 30%-50% in the first year, with cumulative mortalities of 75% and higher in the second year in oysters introduced to an infected area. The disease does not cause serious mortalities below salinities of 12-15 ppt (parts per thousand) but can persist in overwintering oysters in salinities below 5 ppt.

"Dermo" is a warm-temperature disease; outbreaks and mortalities occur in the summer months (June through October in the Chesapeake). Epizootic disease occurs typically in warming temperatures in the 18°C-30°C range.

The disease increases dramatically in infected areas that receive heavy plantings of oysters. It has been shown to be transmitted over at least 50 ft in the water, and it is possible that it is transmissible over much greater distances. The disease can be spread from one oyster to another by *Boonea impressa,* a gastropod parasite of the oyster. *Boonea* can increase the infection intensity of oysters already infected with *Perkinsus* as well as initiate new infections in the oysters on which it feeds.

Diagnosis

The parasites are not visible to the naked eye. As they do with other terminal conditions, oysters infected with dermo will show a weakened shell closure and gape. Decrease in growth is reported to occur several months before the onset of mortalities. Heavy infections of the parasites can be visualized under the microscope with the aid of Lugol's iodine stain (see reference by Ray 1966) as shown in Figure 1. The parasites can be enlarged to facilitate microscopic visualization by culturing the oyster tissues in a specialized microbiological culture medium. This technique and histological confirmation of the disease require professional assistance.

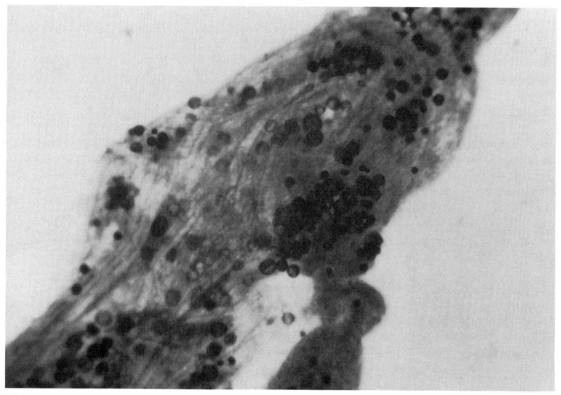

Figure 1. *Perkinsus marinus* parasites, magnified 400 times, in a sample of oyster tissue after five days in a microbiological growth medium (fluid thioglycollate) and stained with iodine. This method can be used to help identify infected oysters.

Prevention and Management
Disease-free Areas
American oysters infected with the disease should not be imported into disease-free areas. Historical records of the disease, in conjunction with the microbiological and histological methods mentioned above, make it possible to be reasonably certain of the presence or absence of the disease in a given population of American oysters.

Although the disease is geographically widespread, there appear to be areas of disease-free American oysters on the Atlantic coast. Conditions on the west coast of North America are not favorable for cultivation of the American oyster, and there is no evidence that the disease affects any of the oysters cultured there. Nevertheless, any proposed movement of American oysters to disease-free areas should include a thorough examination to ascertain the dermo-free status of the oysters.

Areas Known to Have the Disease
Eradication is not considered possible due to the widespread nature of the disease and the lack of knowledge regarding other species that might carry the disease. Management methods consist of reducing the density of oysters and harvesting or moving oysters to low-salinity areas before the warm months.

References

Andrews, J. D., and W. G. Hewatt. 1957. Oyster mortality studies in Virginia. II. The fungus disease caused by *Dermocystidium marinum* in oysters of Chesapeake Bay. *Ecology Monographs* 27:1-26.

Andrews, J. D. 1966. Oyster mortality studies in Virginia. V. Epizootiology of MSX, a protistan pathogen of oysters. *Ecology* 47(1):19-31.

Levine, N. D. 1978. *Perkinsus* gen. n. and other new taxa in the protozoan phylum Apicomplexa. *Parasitology* 64(3):549.

Mackin, J. G., H. M. Owen, and A. Collier. 1950. Preliminary note on the occurrence of a new protistan parasite, *Dermocystidium marinum* n. sp. in *Crassostrea virginica* (Gmelin). *Science* 111:328-329.

Perkins, F. O., and R. W. Menzel. 1966. Morphological and cultural studies of a motile stage in the life cycle of *Dermocystidium marinum*. *Proceedings of the National Shellfisheries Association* 56:23-30.

Ray, S. M. 1952. A culture technique for the diagnosis of infections with *Dermocystidium marinum* Mackin, Owen, and Collier in oysters. *Science* 116:360-361.

Ray, S. M. 1966. A review of the culture method for detecting *Dermocystidium marinum*, with suggested modifications and precautions. *Proceedings of the National Shellfisheries Association* 54:55-69.

White, M. E., E. N. Powell, S. M. Ray, and E. A. Wilson. 1987. Host-to-host transmission of *Perkinsus marinus* in oyster *(Crassostrea virginica)* populations by the ectoparasitic snail *Boonea impressa* (Pyramidellidae). *Journal of Shellfish Research* 6:1-5.

MSX Disease of the American (Eastern) Oyster *(Crassostrea virginica)*

MSX disease is caused by a parasite known as *Haplosporidium nelsoni* (formerly known as *Minchinia nelsoni)* and originally referred to as multinucleate sphere unknown (MSX). The disease is also known as haplosporidiosis of American oysters and Delaware Bay disease. The parasite invades virtually all tissues of the oyster, but apparently it requires another host species (as yet unknown) in order to complete its life cycle.

The disease was first recognized in Delaware Bay in 1957. It rapidly destroyed the Delaware Bay industry, with mortalities of oysters reaching 90%-95% by 1960 (see Figure 2). The percentage of infected oysters and mortality due to the disease have fluctuated over the years. Seed from partially resistant Delaware Bay stock introduced into infected areas are still subject to serious mortalities, reported to be in the 30% range after one year. Evidence shows that this resistance to the disease has occurred in some stocks of oysters subjected to continuing infection over the years (see Figure 3). The parasite causes serious mortalities in Chesapeake Bay as well. Since about 1980, a recurrence of the disease has been observed in both Chesapeake Bay and Delaware Bay, associated with a drought.

Geographic Range and Species Infected

MSX disease has been reported from Maine to Florida, but the most serious mortalities occur in Delaware Bay and Chesapeake Bay. The American oyster is the only oyster known to be infected, but similar parasites infect a variety of other bivalve molluscs.

Mortality Rate, Environmental Factors, and Seasonality

The early infections in Delaware Bay resulted in mortalities approaching 100% over three years. Oyster stocks introduced to Delaware Bay that have not been subjected to MSX disease show similarly high mortality rates. On seed beds in Delaware Bay, annual mortalities of stocks exposed (and thus partially resistant) to the disease were estimated to be in the 4%-9% range, before the drought that began in 1980. About 30% of these seed were estimated to have died within one year when transplanted to infected high-salinity growout beds. Mortalites have been higher in recent years, exceeding 50% even among the resident Delaware Bay stocks.

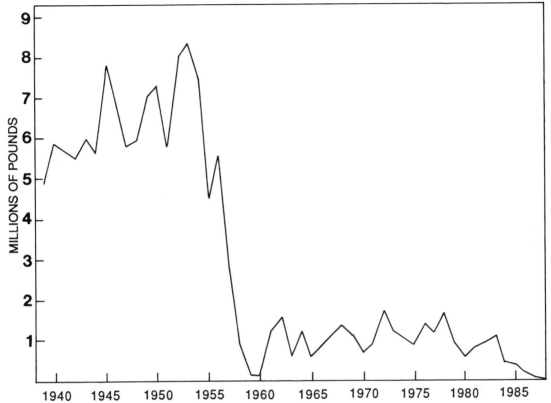

Figure 2. New Jersey oyster production over a fifty-year period. The severe decline in production in the late 1950s corresponds with occurrence of MSX disease in the oyster population. (Graph courtesy S. E. Ford)

Salinity and temperature are known to affect the severity of MSX disease. In general, the disease is rarely acquired below about 10 ppt (parts per thousand); salinities of about 15 ppt are required for the parasite to appear in substantial numbers in host tissues, and serious mortalities occur only above about 20 ppt. There is some indication that the disease may be limited by a salinity greater than 30 ppt.

Oysters become infected during the warm months (late May through October), with peak mortalities in late summer and early fall and again in the following summer. The disease reappears or increases in severity in drought years. The parasite appears to be sensitive to high temperatures; in oysters with some resistance, the disease is reported to go into remission or disappear when temperatures exceed about 20°C.

Diagnosis

A definitive diagnosis requires professional pathological assessment or microscopic examination of oyster tissues. The following signs characterize the disease, but other

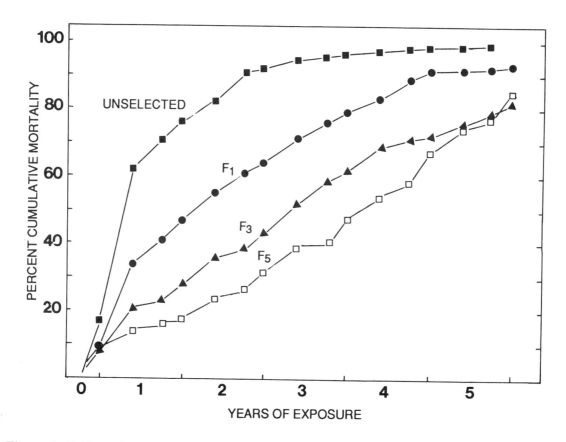

Figure 3. Evidence that oysters from Delaware Bay develop some resistance to *Haplosporidium nelsoni* (the cause of MSX disease) after continuous exposure to the parasite. The highest mortality occurred in the oysters that had not been exposed to MSX. The reduced mortality rates are shown for the first-, third-, and fifth-generation offspring (F_1, F_3, and F_5, respectively) of the original stocks. The graph shows that the greatest differences in mortality rate for the offspring are in their first three years of being exposed to MSX. (Graph courtesy of S. E. Ford, from Ford and Haskin 1987)

diseases may cause similar signs: thin, watery oysters with pale digestive glands; mantle recession and heavy fouling along the interior margins of the left-hand valve in oysters with advanced infections; and, in some cases, raised yellow-brown deposits on the interior valve surfaces in older surviving oysters.

Prevention and Management
Disease-free Areas

As in the case of other infectious oyster diseases that cause serious mortalities, MSX-infected oysters should not be introduced into areas where the disease has not been reported. Historical and pathological evaluation of American oyster stocks should be made to locate disease-free stocks of animals if this species is needed for importation to other areas.

Since the American oyster does not thrive under conditions on the continental west coast of North America, and since MSX disease does not appear to infect the presently cultivated species on that coast, there appears to be little risk from this disease on the west coast today.

Areas Known to Have the Disease

Eradication is not a feasible approach because of the widespread occurrence of the disease and the possible existence of other unknown hosts involved in the life cycle of MSX. In infected areas, it is clear that the disease can be controlled by holding the oysters in low-salinity areas or, where possible, by taking advantage of the sensitivity of the parasite to temperature and salinity. Oysters should be held for as short a time as possible in high-salinity areas where they may be transferred for growing.

For large-scale control of the disease, an MSX-resistant strain of oysters (as developed by researchers at the Rutgers Oyster Research Laboratory in Bivalve, New Jersey) offers the best hope.

References

Andrews, J. D. 1967. Interaction of two diseases of oysters in natural waters. *Proceedings of the National Shellfisheries Association* 57:38-49.

Andrews, J. D., and J. L. Wood. 1967. Oyster mortality studies in Virginia. VI. History and distribution of *Minchinia nelsoni,* a pathogen of oysters in Virginia. *Chesapeake Science* 8(1):1-13.

Farley, C. A. 1965. Acid-fast staining of haplosporidian spores in relation to oyster pathology. *Journal of Invertebrate Pathology* 7:144-147.

Farley, C. A. 1967. A proposed life cycle of *Minchinia nelsoni* (Haplosporida, Haplosporidiidae) in the American oyster *Crassostrea virginica. Journal of Protozoology* 14:616-625.

Farley, C. A. 1975. Epizootic and enzootic aspects of *Minchinia nelsoni* (Haplosporida) disease in Maryland oysters. *Journal of Protozoology* 22(3):418-427.

Ford, S. E. 1985. Effects of salinity on survival of the MSX parasite *Haplosporidium nelsoni* (Haskin, Stauber, and Mackin) in oysters. *Journal of Shellfish Research* 5(2):85-90.

Ford, S. E., and H. H. Haskin. 1982. History and epizootiology of *Haplosporidium nelsoni* (MSX), an oyster pathogen in Delaware Bay 1957-1980. *Journal of Invertebrate Pathology* 40:118-141.

Ford, S. E., and H. H. Haskin. 1987. Infection and mortality patterns in strains of oysters *Crassostrea virginica* selected for resistance to the parasite *Haplosporidium nelsoni* (MSX). *Journal of Parasitology 73(2):368-376.*

Haskin, H. H., and S. E. Ford. 1982. *Haplosporidium nelsoni* (MSX) on Delaware Bay seed oyster beds: A host-parasite relationship along a salinity gradient. *Journal of Invertebrate Pathology* 40:388-405.

Haskin, H. H., L. A. Stauber, and J. A. Mackin. 1966. *Minchinia nelsoni* n. sp. (Haplosporida, Haplosporidiidae): Causative agent of the Delaware Bay oyster epizootic. *Science* 53:1414-1416.

Seaside Haplosporidiosis of American (Eastern) Oysters (*Crassostrea Virginica*)

Seaside haplosporidiosis is caused by a parasite known as *Haplosporidium costale* (formerly *Minchinia costalis).* After its discovery in 1959 it was described as a seaside organism (SSO) due to its occurrence in the more saline waters on the seaside coast of Virginia and Maryland, in contrast to *Haplosporidium nelsoni* (causing MSX disease), which is found in more inland waters such as Chesapeake Bay. The disease caused three years of serious mortalities from 1959 to 1961, but it has not been so severe and recurrent a problem as MSX disease. However, annual mortality rates can reach 50% in seaside bays of Virginia.

The parasite infects all tissues of the oysters except the epithelium and is capable of causing substantial synchronous mortalities when the parasites form spores.

Geographic Range and Species Infected

Only the American oyster is infected. The disease has been reported in some populations from the Virginia coast north to Maine. Serious mortalities from the disease are reported only in high-salinity bays on the seaside coast from Virginia to Delaware.

Mortality Rate, Environmental Factors, and Seasonality

The disease is detectable only by pathological examination from March through June of each year and is associated with mortalities in May and June. Infections acquired in the spring of one year may not cause death until the following spring, when the mortality rate can reach 50%. Oysters may also be infected with MSX disease, which often kills the oyster before it can succumb to *Haplosporidium costale.* Thus, the apparent rate of mortality due to *H. costale* is lower than it would be if the MSX organism were not present.

The fact that this disease occurs on seaside coasts rather than in the more inland embayments apparently results from its requirement for high salinities to infect and cause disease in the host.

Diagnosis

A definitive diagnosis is based on a histological examination of tissues and identification of the parasite; professional assistance is required for the identification of the parasites' life stages. Sick and gaping oysters are thin and may be discolored.

Prevention and Management

Transplantation of oysters to less saline areas retards or eliminates the disease. In the Chesapeake Bay region, this may be effective for seaside haplosporidiosis, but oysters may then become infected with the more serious MSX disease.

References

Andrews, J. D. 1982. Epizootiology of late summer and fall infections of oysters by *Haplosporidium nelsoni,* and comparison to annual life-cycle of *Haplosporidium costale,* a typical haplosporidan. *Journal of Shellfish Research* 2:15-23.

Andrews, J. D., J. L. Wood, and H. D. Hoese. 1962. Oyster mortality studies in Virginia. III. Epizootiology of a disease caused by *Haplosporidium costale,* Wood and Andrews. *Journal of Insect Pathology* 4:327-343.

Andrews, J. D., and M. Castagna. 1978. Epizootiology of *Minchinia costalis* in susceptible oysters in seaside bays of Virginia's eastern shore, 1959-1976. *Journal of Invertebrate Pathology* 32:124-138.

Velar Virus Disease of Pacific Oysters

Oyster velar virus disease (OVVD) is known only as a hatchery disease and is commonly referred to as "blisters" by hatchery workers. The virus causing the disease belongs to a group known as the iridoviruses. It infects the epithelium of the velum of the larva and can cause serious mortalities in hatchery operations. The virus is most likely carried in the adult oyster and may even cause some form of disease in the adult, but this has not been documented.

Geographic Range and Species Infected

Washington is the only state that has reported the presence of OVVD. However, considering the historical commerce of this oyster around the Pacific Rim, it is likely that the disease is much more widespread than is now known.

Larvae of the Pacific oyster, *Crassostrea gigas,* are the only species and life stage known to be infected by the disease. Although the disease has been confirmed only in hatchery-reared larvae, it probably occurs in some wild stocks as well. Similar viruses have been observed in adult Pacific and Portuguese oysters in France, but their relationship to OVVD has not been determined.

Mortality Rate, Environmental Factors, and Seasonality

Oyster velar virus disease can cause nearly 100% mortality in affected hatchery tanks. It is not known which environmental factors affect the disease or the susceptibility of the oysters. The disease typically appears from March to May, but it has also been reported throughout the summer.

Diagnosis

OVVD can be suspected when mortalities occur in the spring, always in larvae greater than 150 µm in shell length and at least 10 days after spawning, when grown in the 25°C-30°C temperature range. Virus-infected cells on the velum of sick larvae detach and form the characteristic "blisters." The larval velums also lose their cilia. The blisters and loss of cilia can be observed easily with the aid of a microscope (see Figures 4 and 5).

It should be noted that loss of cilia can result from other causes such as storing larvae at too high or too low a temperature during shipment to a remote setting site. A shellfish pathologist can confirm the diagnosis by examining the larval tissues for typical lesions caused by the virus.

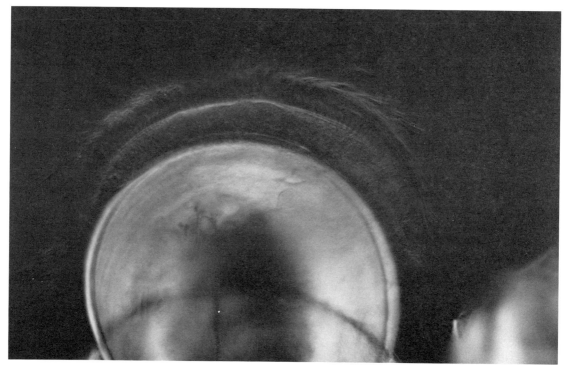

Figure 4. Normal Pacific oyster larva about 10 days of age, with a shell dimension of about 0.2 mm, showing the normal extended velum with cilia. (From Elston and Wilkinson 1985)

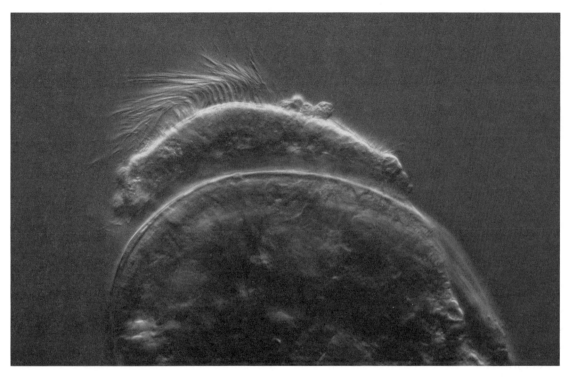

Figure 5. Pacific oyster larva with oyster velar virus disease (OVVD). It has lost cilia from the velum and has formed "blisters," which are cells of the oyster infected with the virus. Larvae in this condition would be found on the tank bottom, but larvae in an earlier stage of the disease can be found swimming in the water column. (From Elston and Wilkinson 1985)

Prevention and Management
Disease-free Areas
Although it is likely that OVVD is more widely distributed on the Pacific Rim than is known today, larvae infected with the virus should not be introduced into areas not known to have the disease.
Areas Known to Have the Disease
Eradication is not currently possible. If the source of the virus that infects larvae in the spring spawning season is determined to be the adult oysters, it may eventually be possible to identify virus-free brood stock and eliminate the disease.

Effective management can be practiced if hatchery personnel recognize the infectious nature of the disease. If larval tanks are suspected of being contaminated, steps should be taken to avoid contaminating other tanks in the hatchery. These steps include the sterilization of screens and other equipment by rinsing them for 15 minutes in a solution of 10 parts per million sodium hypochlorite or other disinfectant between uses on different tanks, and careful personnel procedures to prevent cross contamination. If the disease is confirmed, affected tanks should be sterilized immediately, larvae discarded, and rigorous procedures instituted to prevent tank-to-tank spread.

The steps used in the hatchery to control the spread of OVVD and other diseases are summarized elsewhere in the guide.

References

Elston, R. 1979. Virus-like particles associated with lesions in larval Pacific oysters *(Crassostrea gigas)*. *Journal of Invertebrate Pathology* 33:71-74.

Elston, R. A., and M. T. Wilkinson. 1985. Pathology, management and diagnosis of oyster velar virus disease (OVVD). *Aquaculture* 48:189-210.

Denman Island disease of the Pacific Oyster *(Crassostrea gigas)*

Denman Island disease is a little understood but important malady of the Pacific oyster. It is apparently caused by a parasite that lives within the glycogen storage cells of the oyster and is now known as *Mikrocytos mackini*. Little is known about the biology of this parasite, but affected oysters may die at a high rate and surviving oysters do not ripen for spawning.

Denman Island disease is sometimes called "microcell" disease. This term has also been used to refer to bonamiasis (caused by *Bonamia ostreae)* of the European flat oyster. Denman Island disease and bonamiasis are two different diseases of two different oysters. Since the diseases and their causative microorganisms appear to be unrelated, the term "microcell" should be abandoned in both cases to avoid further confusion between these two diseases.

Geographic Range and Species Infected

The disease was first reported in Pacific oysters, *Crassostrea gigas,* from Henry Bay on Denman Island in British Columbia, Canada, in 1960. Since then it has been noted at other sites around Denman Island and at other locales in the Strait of Georgia in British Columbia. It is only known to infect the Pacific oyster.

Mortality Rates, Environmental Factors, and Seasonality

Mortality rates up to 53% in a single season have been reported, but the severity fluctuates from year to year. Infection and loss to the disease increased at lower tide levels when oyster mortality was monitored at the 4.0, 2.5, and 1.0 ft levels. In the original report of the disease, only oysters from five to seven years old were affected; two-year-old oysters appeared to be healthy.

Signs of the disease can appear in April and develop through July. Mortality losses occur by July and the disease is considered to have run its course by August. Some of the infected oysters recover.

Diagnosis

The disease is characterized by the appearance of round yellow-to-green lesions or pustules (1-3 mm in diameter) on the body surface. Since similar lesions occur in several other oyster diseases, definitive diagnosis should be made by a pathologist's microscopic examination.

Prevention and Management

Disease-free Areas

Since the disease is infectious and seems to be confined in its geographic distribution, infected oysters should be moved only into areas where the disease is known to occur already.

Areas Known to Have the Disease

Eradication of this disease is not feasible. The geographic distribution of Denman Island disease, while not precisely known, is too large to make eradication feasible. The only management advice that can be drawn from the little information available on the disease would be to move oysters above the 2.5 ft tidal level during the period in which the disease is active, since the disease occurs at a lower rate at the higher tide levels. Presumably, the effects of the disease could be reduced by avoiding planting at tide levels below about 2 ft before June.

References

Bower, S. M. 1989. Circumvention of mortalities caused by Denman Island disease during mariculture of *Crassostrea gigas.* In *Disease Processes in Marine Bivalves,* ed. W. F. Fisher, American Fisheries Society Special Publication 18, Washington, D. C.

Farley, C. A., P. H. Wolf, and R. A. Elston. 1988. A long-term study of "microcell" disease in oysters with a description of a new genus, *Mikrocytos* (G. N.), and two new species, *Mikrocytos mackini* (Sp. N.) and *Mikrocytos roughleyi* (Sp. N.). *Fishery Bulletin* (U.S.) 86:581-593.

Quayle, D. B. 1982. Denman Island oyster disease 1960-1980. *Shellfish Mariculture Newsletter* 2(2):1-5.

Quayle, D. B. 1961. *Denman Island oyster disease and mortality, 1960.* Manuscript No. 713, Report Series (Biological), Fisheries Research Board of Canada, 6 pp.

Nocardiosis of the Pacific Oyster

Nocardiosis is a bacterial disease of the Pacific oyster, *Crassostrea gigas.* This name is new, a result of the recent isolation of the bacterium belonging to the group *Nocardia* (see reference by Friedman). The disease has previously been known as "fatal inflammatory bacteraemia," "focal necrosis," and "multiple abscesses."

It is likely that nocardiosis is an important component of the phenomenon known in the Pacific Northwest as "summer mortality." The disease causes typical small, round yellow lesions on the body surfaces of the oysters, often observed on gaping individuals. The lesions are abcesses caused by the bacteria, which are distributed throughout the oyster's body in its blood system.

Geographic Range and Species Infected

A disease described as "multiple abscesses" from Matsushima Bay, Japan, appears to be the same as nocardiosis. On the west coast of North America, the disease has been reported from sites in California, including Tomales Bay; in Washington from Willapa Bay and south Puget Sound embayments; and at several sites in British Columbia including Nanoose Bay, Scott Island, and Ships Point. It is likely that the disease is more geographically widespread than these observations indicate.

The Pacific oyster, *Crassostrea gigas,* is the principal oyster affected by the disease, although a few specimens of the European flat oyster, *Ostrea edulis,* cultivated near areas of infected Pacific oysters have been reported to have a similar disease.

Mortality Rate, Environmental Factors, and Seasonality

The mortality rate due to nocardiosis has not been accurately measured. However, the severity of the disease in individual oysters and the high prevalence in some populations suggest that it is a significant mortality factor. In one study it was reported to occur in about 30% of oysters sampled from sites in south Puget Sound, Washington, during September and October. The widespread geographic occurrence of the disease suggests that it can occur wherever the Pacific oyster is cultured and that the causative bacterium is ubiquitous and acquired from the environment.

In Puget Sound, shallow bays that are subject to warm summer temperatures appear to have the most frequent and severe occurrence of the disease. In British Columbia, the disease has been found in other areas beside warm shallow embayments and on firm, rocky and sand bottoms. The disease is found in Puget Sound in oysters from mud-bottom bays but has also been reported on gravel bottoms. The disease is a summer and fall phenomenon, typically observed from August through November. The disease may be present in some populations at other times of the year but at a lesser intensity.

Diagnosis

Round yellow-to-green lesions often about 2 mm but up to 1 cm in diameter on the surface of the mantle or gill (as shown in Figure 6), and often in the area of the adductor

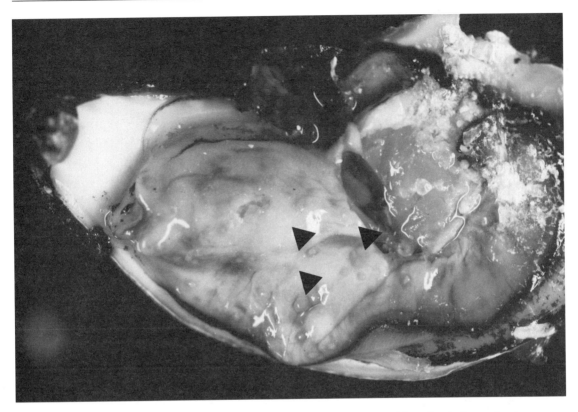

Figure 6. Adult Pacific oyster with nocardiosis. This disease is caused by a bacterium whose presence produces yellow spots on the body surface (arrows) shown in the photograph. Other diseases can cause similar discolorations and raised spots on the body surface. (From Elston et al. 1987)

muscle and heart, are indicative of nocardiosis. Because other diseases, such as Denman Island disease, produce similar lesions, oysters should be submitted to a pathologist for microscopic examination of the tissues for a definitive diagnosis.

Prevention and Management
Disease-free Areas
The true geographic distribution of nocardiosis is not known. However, it appears to be widespread and the causative organism may be acquired from the environment. Given the historical movements of the Pacific oyster around the Pacific Rim, no areas should be assumed to be disease-free. While it is not good husbandry to transplant obviously infected oysters, the condition cannot be considered an exotic disease to the west coast of North America, and its transfer within this area does not appear to present any unusual risk.

Areas Known to Have the Disease
For the reasons noted above, eradication of nocardiosis is not feasible. Management

techniques have not been tested, but it is possible that culturing oysters in off-bottom systems will reduce the prevalence and severity of the disease. Where possible, moving oyster stocks out of warm shallow embayments by harvest or transfer to other growout areas could reduce the impact of this disease.

References

Elston, R. A., J. H. Beattie, C. Friedman, R. Hedrick, and M. L. Kent. 1987. Pathology and significance of fatal inflammatory bacteraemia in the Pacific oyster, *Crassostrea gigas* Thünberg. *Journal of Fish Diseases* 10:121-132.

Friedman, C. S., B. L. Beaman, R. P. Hedrick, J. H. Beattie, and R. A. Elston. 1988. Nocardiosis of adult Pacific oysters, *Crassostrea gigas*. *Journal of Shellfish Research* 7(1):216.

Glude, J. B. 1974. A summary report of the Pacific coast oyster mortality investigations 1965-1972. Proceedings of the Third U.S.-Japan Meeting on Aquaculture at Tokyo, Japan, October 15-16, 1974.

Imai, T., K. Numachi, J. Oizumi, and S. Sato. 1965. Studies on the mass mortality of the oyster in Matsushima Bay. II. Search for the cause of mass mortality and the possibility to prevent it by transplantation experiment. (In Japanese, English summary.) *Bulletin of Tohoku Regional Fisheries Research Laboratory* 25:27-38.

Imai, T., K. Mori, Y. Sugawara, H. Tamate, J. Oizumi, and O. Itakawa. 1968. Studies on the mass mortality of oysters in Matsushima Bay VII. Pathogenetic investigation. *Tohoku Journal of Agricultural Research* 19:250-257.

Lauckner, G. 1983. *Diseases of Marine Animals,* vol. 2, *Introduction. Bivalvia to Scaphopoda.* Biologische Anstalt Helgoland, Hamburg.

Sindermann, C. J., and A. Rosenfield. 1967. Principal diseases of commercially important marine bivalve mollusca and crustacea. *Fishery Bulletin* (U.S.) 66:335-385.

Bonamiasis of the European Flat Oyster
(Ostrea edulis)

The disease known as bonamiasis is caused by a parasite *(Bonamia ostreae)* that infects the blood cells of the oyster. The parasite eventually infects all of the oyster's blood cells, destroying its "immune" system and interfering with other physiological processes. Serious mortalities (up to almost 100% per year) can occur in newly infected populations. It is transmitted by water contact, but close proximity to infected oysters is believed to be necessary.

It is now known that bonamiasis occurred in flat oysters in California in the 1960s, but it was then known as "microcell disease." The disease spread from a California hatchery to Brittany, France, where it initiated the well-known epizootic. The disease is best known for its substantial impact on the European industry, particularly in France (see Figure 7), where it was first identified in 1979. Bonamiasis was transplanted to Washington from the California hatchery in the late 1970s and remains an important disease in the Pacific Northwest.

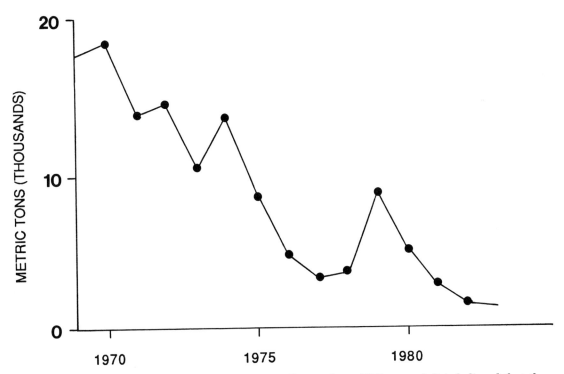

Figure 7. Production of the European flat oyster in France from 1970 onward. It is believed that the first drop in production, up to about 1977, was due to marteiliasis and that the drop in production from about 1978 onward resulted, at least in part, from the occurrence of bonamiasis. (Graph courtesy of D. J. Alderman)

Geographic Range and Species Infected

In Europe, bonamiasis is found on the Atlantic coast of France and Spain, in the Netherlands, England, and Ireland. It has recently been discovered in the Mediterranean, but it does not appear to cause serious oyster mortalities there. In North America, the disease is known to occur in California and Washington. *Ostrea edulis* is the primary commercial species affected, but research has shown that the Chilean oyster *(Ostrea chilensis)* and the New Zealand dredge oyster *(Tiostrea lutaria)* can also be infected with *Bonamia ostreae*. Experiments suggest that the Olympia oyster *(Ostrea lurida)* may contract the disease, but this has not been positively demonstrated. A similar but distinctive parasite occurs only in New Zealand in the dredge oyster.

Mortality Rate, Environmental Factors, and Seasonality

High mortality rates approaching 100% within one year are reported in newly infected populations in Europe. Mortality rates tend to decline the longer a population has been infected and are typically reported in the 20%-60% range after the disease has been in a population for two or more years. In populations with a long history of the disease, mortality tends to be highest in younger oysters, with the annual mortality rate declining as the oysters get older. For example, studies in Washington showed 20%, 7%, and 4% mortality rates, respectively, for two-, three-, and four-year-old infected oysters.

The disease can appear throughout the year, but it is generally associated with warming spring and summer temperatures. Bonamiasis can cause significant mortalities between 12°C and 20°C but not at higher temperatures.

Diagnosis

No specific signs of bonamiasis can be detected with the naked eye. Some oysters may show a weakened shell closure and gape when the disease is in an early stage, but in other cases these signs do not occur until the disease is advanced. Confirmatory diagnosis requires professional assistance and the use of microscopic and antibody techniques. Microscopic examination will reveal the distinctive parasite within the blood cells of the oyster (Figure 8). A field diagnostic kit produced in France is now available to shellfish farmers.

Prevention and Management

Disease-free Areas

In areas where the disease does not occur, the best strategy is to ensure that infected oysters are not introduced. A thorough historical and pathological examination by a professional should be performed on any lot of flat oysters before they are introduced to such an area. If infected animals are introduced into a population with no prior exposure to the disease, high mortalities can be expected for at least five years.

Areas Known to Have the Disease

Experiments in the Netherlands indicate that it may be possible to eradicate bonamiasis by completely removing flat oysters for a minimum of three years. Eradication is a reasonable alternative only if the oyster grounds can be used for culture of another non-susceptible species such as *Crassostrea gigas*. Before flat oysters are reintroduced into the

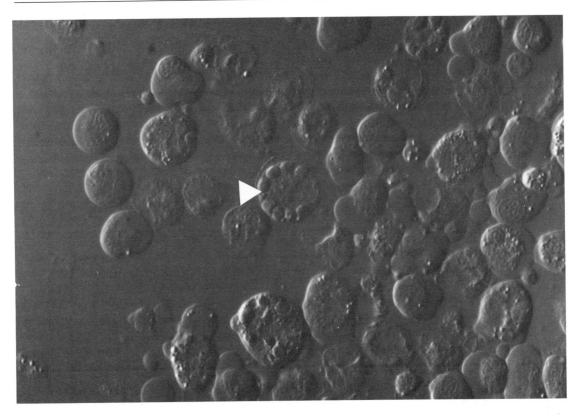

Figure 8. European flat oyster blood cells, highly magnified. The cell in the center of the photograph (arrow) is infected with *Bonamia*, the smaller round spheres within the blood cell. (From Elston et al. 1986)

area on a commercial scale, test batches should be introduced and examined over a two-year period. Thus, the total eradication period will take at least five years.

If economic or other practical considerations prevent the eradication approach, some steps can be taken to reduce the effects of the disease. It is known that mortalities due to bonamiasis are reduced in off-bottom culture methods.

Reducing the density of oysters is also believed to reduce the transmission of the disease, as is the use of subtidal rather than intertidal growing areas. In addition, it appears that some stocks of flat oysters may acquire resistance to the disease. These populations of oysters still carry the infectious parasite and some individual oysters succumb to the disease, but many appear to tolerate and grow well in spite of the infection.

Infected oyster populations should not be used as brood stock for seed to be planted into disease-free areas. There is no reason, however, to avoid the introduction of infected oyster seed into areas known to be infected and in which eradication is not possible.

References

Balouet, G., J. Poder, and A. Cahour. 1983. Haemocytic parasitosis: Morphology and pathology of lesions in the French flat oyster, *Ostrea edulis* L. *Aquaculture* 34:1-14.

Bucke, D., and S. Feist. 1985. Bonamiasis in the flat oyster, *Ostrea edulis,* with comments on histological techniques. In *Fish and Shellfish Pathology,* ed. A. E. Ellis (Academic Press, London), pp. 387-392.

Comps, M., G. Tigé, H. Grizel, and C. Vago. 1980. Etude ultrastructural d'un prostiste parasite de l'Huître *Ostrea edulis* L. *Comptes Rendus Académie des Sciences Paris* 290, Série D: 383-384.

Elston, R. A., C. A. Farley, and M. L. Kent. 1986. Occurrence and significance of bonamiasis in European flat oysters *Ostrea edulis* in North America. *Diseases of Aquatic Organisms* 2:49-54.

Elston, R. A., M. L. Kent, and M. T. Wilkinson. 1987. Resistance of *Ostrea edulis* to *Bonamia ostreae* infection. *Aquaculture* 64:237-242.

Holsinger, L. M. 1988. *Bonamia ostreae:* the protozoan parasite in Washington stocks of *Ostrea edulis* and the influence of temperature on disease development. Master's thesis, University of Washington, Seattle.

Katkansky, S. C., and R. W. Warner. 1974. *Pacific oyster disease and mortality studies in California.* Marine Resources Technical Report No. 25, California Department of Fish and Game, Long Beach, 51 pp.

Katkansky, S. C., W. A. Dahlstrom, and R. W. Warner. 1969. Observations on survival and growth of the European flat oyster, *Ostrea edulis,* in California. *California Fish and Game* 55:69-74.

Pichot, Y., M. Comps, G. Tigé, H. Grizel, and M.-A. Rabouin. 1980. Recherches sur *Bonamia ostreae* gen n., sp. n., parasite nouveau de l'huître plate *Ostrea edulis* L. *Revue des Travaux de l'Institut des Pêches Maritimes* 43:131-140.

Poder, M., A. Cahour, and G. Balouet. 1982. Hemocytic parasitosis in European oyster *Ostrea edulis* L.: pathology and contamination. In *Invertebrate Pathology and Microbial Control,* ed. C. C. Payne and H. D. Burges (Society for Invertebrate Pathology, Brighton, U.K.), pp. 254-257.

Van Banning, P. 1982. Some aspects of the occurrence, importance and control of the oyster pathogen *Bonamia ostreae* in Dutch oyster culture. In *Invertebrate Pathology and Microbial Control,* ed. C. C. Payne and H. D. Burges (Society for Invertebrate Pathology, Brighton, U.K.), pp. 261-265.

Van Banning, P. 1985. Control of *Bonamia* in Dutch oyster culture. In *Fish and Shellfish Pathology,* ed. A. E. Ellis (Academic Press, London), pp. 393-396.

Marteiliasis of the European Flat Oyster (*Ostrea edulis*)

Marteiliasis (sometimes called Aber disease) is caused by a parasite, *Marteilia refringens,* that infects the connective and digestive tissues of the oyster. Spores (the mature stage of the parasite) are formed in the epithelium of the digestive tubules. The disease is responsible for flat oyster mortalities that began in 1967 along certain regions of Atlantic France and Spain.

A related parasite of the Australian rock oyster, *Saccostrea commercialis,* is *Marteilia sydneyi.* This parasite has caused heavy mortalities of the rock oyster in Moreton Bay, Queensland, Australia.

Geographic Range and Species Infected

Marteiliasis occurs only on the Atlantic coast of Europe. Serious disease resulting from the parasite infection, first reported from Aber Wrach in Brittany in 1967, occurs in other areas of France and in Spain as well. *Marteilia* parasites have been observed in Dutch flat oysters but without significant disease or mortality.

The disease occurs only in *Ostrea edulis,* but the parasite has been found, according to a single report from France, in a few specimens of the Pacific oyster, *Crassostrea gigas.* No significant detriment to health was reported in the Pacific oyster as a result of the infection, but the identity of the parasites observed needs to be confirmed as *Marteilia refringens* before accepting this as a definitive observation.

Mortality Rate, Environmental Factors, and Seasonality

Mortality rates of 90% annually were reported in the first epizootics of disease in France. When disease-free spat or two- and three-year-old oysters were planted in infected areas in March, they became infected between the first of May and the end of August. Severe mortalities occurred before the end of the first winter, but the parasite could not be found in the surviving oysters. The fact that the parasite occurs in oysters in some areas without causing disease suggests that environmental factors or oyster stock differences are important in determining whether or not the disease becomes a significant problem.

In addition, mortality seems to be related to the formation of the spore stages (known as sporulation) of the parasite within the oyster tissues. The sporulation process may result in the release of toxic substances that affect the oyster.

Diagnosis

Heavily infected oysters may have normally dark colored digestive glands and abundant glycogen stored in the connective tissues. In some cases, however, in diseased oysters with advanced infections, the mantle is colorless, the digestive gland is pale yellow rather than brown, and the visceral mass is slimy and shrunken. Definitive diagnosis requires the histological examination of the tissues by a professional pathologist.

Prevention and Management
Disease-free Areas
Infected oysters should not be moved into disease-free areas.
Areas Known to Have the Disease
No management methods are known for the disease.

References
Balouet, G. 1979. *Marteilia refringens:* Considerations of the life cycle and development of Abers disease in *Ostrea edulis. Marine Fisheries Review* 14(1-2):45-53.

Cahour, A. 1979. *Marteilia refringens* and *Crassostrea gigas. Marine Fisheries Review* 41(1-2):19-20.

Grizel, H. 1979. *Marteilia refringens* and oyster disease: recent observations. *Marine Fisheries Review* 41(1-2):3-7.

Grizel, H., M. Comps, F. Cousserans, J.-R. Bonami, and C. Vago. 1974. Pathologie des invertébrés. Etude d'un parasite de la glande digestive observé au cours de l'épizöotie actuelle de l'huître plate. *Comptes Rendus Académie des Sciences Paris* 279, Série D:783-784.

Perkins, F. O. 1976. Ultrastructure of sporulation in the European flat oyster pathogen, *Marteilia refringens:* taxonomic implications. *Journal of Protozoology* 23(1):64-74.

Gill Disease of the Portugese Oyster (*Crassostrea angulata*)

Gill disease is a term used to describe the condition of Portuguese oysters *(Crassostrea angulata)* subject to a massive mortality which eventually resulted in the elimination of the culture of this oyster from all parts of France and Portugal. The cause is not certain, but several infectious agents have been proposed, among them parasites and a virus. Severe losses were reported in 1967-68 and 1970-73 in the Ile de Oleron and Archachon regions of France and in regions of Portugal.

Geographic Range and Species Infected
Gill disease in the Portuguese oyster has been reported in France, Portugal, and Great Britain. A disease with similar signs has been reported in the European flat oyster,

Ostrea edulis, in several European countries, but it has not been identified definitely as the same disease as afflicts the Portuguese oyster.

Mortality Rate, Environmental Factors, and Seasonality

Because of the devastating effect of gill disease, the Portuguese oyster is no longer cultured in the Ile de Oleron, a major production area for oysters in France. That region now cultures the Pacific oyster, *Crassostrea gigas.*

When Portuguese oysters were imported to Great Britain, it was reported that within three weeks the percentage of oysters showing the "active" disease, in which the gills eroded and were found to contain dead tissue, increased from 2% to 60%. This active stage of the disease was found primarily in spring and summer.

Diagnosis

A preliminary diagnosis can be made on the basis of visible signs. The disease first appears as yellow spots on the gills. As these spots enlarge, the centers become brown and necrotic, resulting in a perforation of the gill, or a V-shaped indentation if the lesion occurs at the edge of the gill. Yellow or green pustules may appear on the adductor muscle or mantle; on the mantle they may develop into perforations. These perforations or indentations of the gill may be found in recovering oysters, but the lesions lack the decaying yellow and brown tissue typical of the active stage of the disease.

Diagnosis can be confirmed by a shellfish pathologist. However, as noted above, the exact cause of the disease has not been determined, although the gills of some oysters with lesions contain a virus.

Prevention and Management

Little is known about the prevention and management of the disease. However, since there is some evidence that a virus or other infectious agent causes the disease, it is not advisable to move oyster stocks known to have had the disease to areas where the disease has not been reported.

References

Alderman, D. J., and P. Gras. 1969. "Gill Disease" of Portuguese oysters. *Nature* 224:616-617.

Comps, M. 1969. Observations relatives a l'affection branchiale des huîtres portugaises *(Crassostrea angulata* Lmk.). *Revue des Travaux de l'Institut des Pêches Maritimes* 33(2):151-160.

Comps, M. 1980. Mise en évidence par fluorescence du virus de la maladie des branchies de l'huître portugaise *Crassostrea angulata* Lmk. *Science et Pêches, Bulletin Institut Pêches Maritimes* 301:17-18.

Comps, M., J. R. Bonami, and C. Vago. 1976. Pathologie des invertébrés: une virose de l'huître portugaise *(Crassostrea angulata* Lmk.). *Comptes Rendus Académie des Sciences Paris* 282, Série D (22):1991-1993.

Franc, A., M. Arvy, and P. P. Gras. 1969. Biologie: Sur *Thanatostrea polymorpha* n.g., n.sp., agent de destruction des branchies et des palpes de l'Huître portugaise. *Comptes Rendus Académie des Sciences Paris* 268, Série D:3189-3190.

Marteil, L. 1969. La maladie des branchies des huîtres portugaises des côtes françaises de l'atlantique. *Revue des Travaux de l'Institut des Pêches Maritimes* 33(2):145-150.

Hexamitiasis of *Ostrea* and *Crassostrea* Oysters

Hexamitiasis is caused by a parasite known as *Hexamita nelsoni.* The disease is also known as "pit disease," a name derived from the belief that it has been responsible for flat oyster mortalites in recirculating water basins, or pits, in Holland. The parasite is considered to be cosmopolitan, that is, to occur commonly throughout the world under suitable conditions.

The parasite is often found in the blood stream and within blood cells in dying oysters, and there is some controversy as to whether it actually causes a disease or simply takes advantage of an oyster already sick from some other cause. The only oyster for which the true disease-causing nature of the parasite has been shown is the Olympia oyster, *Ostrea lurida,* although hexamitiasis has been reported in several other species.

Geographic Range and Species Infected

As noted above, a true disease-causing relationship to the oyster has been established only in *Ostrea lurida* in Puget Sound, Washington. Other species and locations of infection have been reported as follows: *Crassostrea commercialis* (Australian rock oyster), Australia; *Crassostrea gigas* (Pacific oyster), Pacific Northwestern United States; *Crassostrea virginica* (American oyster), Prince Edward Island, Canada; and *Ostrea edulis* (European flat oyster), Holland and the maritime provinces of Canada.

Mortality Rate, Environmental Factors, and Seasonality

Mortality rates have not been recorded precisely, but in certain years oyster farmers have estimated mortalities of about 75% over a two-month period in association with hexamitiasis in *Ostrea lurida.* This is definitely a cold-temperature disease in this species. Experiments show that infection and debilitating disease occur at 6°C and lower but not at 12°C or higher. Mortalities associated with this disease are usually reported in winter, but in Alaska and other northern zones the disease has been found at other times of the year as well.

Diagnosis

A preliminary diagnosis can be made microscopically on a drop of oyster blood. The causative organisms are highly motile by means of their flagella. Confirmation by examining tissues must be made by a shellfish pathologist.

Prevention and Management

Since the causative organism is considered to be cosmopolitan, any oyster-growing area is potentially subject to hexamitiasis. In one of the original publications on the disease in *Ostrea edulis* in Holland (see reference by Mackin et al. 1952), it was suggested that cold temperature, poor circulation over the oyster basins, and overcrowding are optimal conditions for an outbreak of the disease. This is the only published information that provides any hint toward disease management of hexamitiasis.

References

Feng, S. Y., and L. A. Stauber. 1968. Experimental hexamitiasis in the oyster *Crassostrea virginica*. *Journal of Invertebrate Pathology* 10:94-110.

Mackin, J. G., P. Korringa, and S. H. Hopkins. 1952. Hexamitiasis of *Ostrea edulis* L. and *Crassostrea virginica* (Gmelin). *Bulletin of Marine Science of the Gulf and Caribbean* 1:266-277.

Scheltema, R. S. 1962. The relationship between the flagellate protozoan *Hexamita* and the oyster *Crassostrea virginica*. *Journal of Parasitology* 48:137-141.

Schlicht, F. G., and J. G. Mackin. 1968. *Hexamita nelsoni* sp. n. (Polymastigina: Hexamitidae) parasitic in oysters. *Journal of Invertebrate Pathology* 11:35-39.

Stein, J. E., J. G. Denison, and J. G. Mackin. 1959. *Hexamita* sp. and an infectious disease in the commercial oyster *Ostrea lurida*. *Proceedings of the National Shellfisheries Association* 50:67-81.

Shell Disease of Oysters

Shell disease, first described in 1894, is caused by a fungus known as *Ostracoblabe implexa*. Technically, the disease is known as oyster ostracoblabiasis in reference to the fungus. Serious mortalities are thought to have resulted from the disease in Europe at various periods during the 19th and 20th centuries. The disease has been known as *maladie du pied* ("disease of the foot," even though the adult oyster does not have a foot) and *maladie de la charnière* ("disease of the hinge ligament").

The filamentous fungus grows through the shell, weakening it and causing dark raised warts on the interior shell surface. In advanced cases, warts occur in the hinge region and cause excessive and abnormal hinge development. The result may be a beaked appearance to the hinge area of the shell and malformed valves that do not close properly.

Geographic Range and Species Infected

The full disease syndrome, including the formation of warts, occurs only in the European flat oyster *(Ostrea edulis)* in the Netherlands, France, Great Britain, and Nova

Scotia in North America. Part of the disease syndrome, not usually including the wart stages, occurs in *Crassostrea* species. In the Portuguese oyster *(Crassostrea angulata)* a form of shell disease is reported in the Netherlands, France, and Great Britain. Recently the disease was reported in *Crassostrea cucullata* in India. There are reports that a similar disease occurs in the American oyster *(Crassostrea virginica)* on the Atlantic seaboard of North America and in *Crassostrea gryphoides* in India, but these are not confirmed to be shell disease. The fungus that causes the disease is probably common to all marine coastal environments.

Mortality Rate, Environmental Factors, and Seasonality

Shell disease may have caused massive mortalities of *Ostrea edulis* in the Netherlands at various times and has also been claimed to be associated with severe oyster kills in France. Definitive proof that the disease is responsible for the oyster kills is lacking. Oysters are infected above 20°C, either by a waterborne fungus or by direct growth of the fungus from one oyster to adjacent oysters. Young oysters are reported to be more susceptible than older oysters. In the Netherlands, cockle shells, used as spat collectors, were suspected of containing the disease-causing fungus, ensuring that new oyster spat would become infected at an early age.

Diagnosis

A strong presumptive or probable diagnosis can be made on the basis of lesions on the oyster shell. The initial stage of shell disease in one-year-old oysters is the occurrence of small, bright white spots in the growing margin of the shell. This early stage can be cured by chemical treatment, but not the later stages characterized by the "warts" described below.

As the disease progresses, white spots from 0.5 to 3.0 mm in diameter occur on the inner surface of the shell. These spots form a small, slightly raised rough area. A dark indentation in the center of the area indicates that the fungus has penetrated into the mantle cavity. These infected spots coalesce to form the typical "cloud," also with a characteristic rough surface, as the infected area of the shell matrix enlarges. The pallial surface of the shell may acquire a brownish tint in advanced infections.

Formation of "warts" is common. These consist of small green to black protrusions attached to the inner shell surface, often in the area of the adductor muscle attachment and the hinge region but also at other sites on the inner shell surface. Excessive and abnormal hinge deposition may occur and result in a beaked appearance of the dorsal region and inability to effect normal shell closure.

Diagnosis can be confirmed by microscopic examination of the warts and weakened infected shell for the typical forms of the fungus. Fresh shell material should be submitted to the pathologist or preserved for later examination.

Prevention and Management
Disease-free Areas
Since the geographic extent of the disease is not known for certain, it is advisable not to import shell or live oysters from areas known to have the disease into areas where the disease is not known to occur.

Areas Known to Have the Disease
The disease was controlled in the Netherlands by dipping the oysters in a solution of mercuric chloride. However, given our current knowledge of mercury toxicity, this method should not be attempted. It is likely that other methods of killing the fungus would also be effective, such as dipping the oysters in a solution of 15 parts per million sodium hypochlorite (bleach) for 10 minutes or longer. This concentration is made by diluting household bleach containing 5.25% sodium hypochlorite by a factor of 3,500. In the Netherlands, old shell was removed from the oyster beds in order to eliminate a source of the fungus, and areas where young oysters are placed were kept free of dead shells in order to limit the effects of the disease.

References

Alderman, D. J., and E. B. G. Jones. 1971. Physiological requirements of two marine phycomycetes, *Althornia crouchii* and *Ostracoblabe implexa. Transactions of the British Mycological Society* 57:213-225.

Alderman, D. J., and E. B. G. Jones. 1971. *Shell disease of oysters.* Ministry of Agriculture, Fisheries and Food. Fish Investigations Series II, 26(8), 19 pp.

Durve, U. S., and D. V. Bal. 1960. Shell disease in *Crassostrea gryphoides* (Schlotheim). *Current Science* 29:489-490.

Korringa, P. 1947. Les vicissitudes de l'ostréiculture Hollandaise élucidées par la science ostréicole moderne. *Ostréiculture, Cultures Marines* 16(3):3-9.

Korringa, P. 1948. Shell disease in *Ostrea edulis:* its dangers, its cause, its control. National Shellfisheries Association, 1947 Convention Addresses, pp. 89-94.

Korringa, P. 1951. Investigations on shell disease in the oyster *Ostrea edulis* L. *Rapports et Procès-Verbaux des Réunions Conseil International pour l'Exploration de la Mer* 128(2):50-54.

Korringa, P. 1952. Recent advances in oyster biology. *Quarterly Review of Biology* 27:266-308, 339-365.

Raghukumar, C., and V. Lande. 1988. Shell disease of the rock oyster, *Crassostrea cucullata* Born, from Goa, caused by fungus. *Diseases of Aquatic Organisms* 4:77-81.

OTHER DISEASES, OTHER MOLLUSCS

Hemic Neoplasia of Bivalve Molluscs

The disease known as hemic, hematopoietic, or hemocytic neoplasia (HCN) is also referred to as hemic proliferative disease, leukocytic neoplasia, sarcomatous neoplasia, sarcomatoid proliferative disorder, disseminated sarcoma, and atypical hemocyte condition. As a neoplasia, it is considered to be a form of cancer of shellfish similar to leukemia in higher animals and man in the way it affects the host. It should be emphasized, however, that this is a cancer of shellfish, not of humans, and that consuming shellfish with this condition poses no known health threat to humans.

Some research has suggested that the disease is caused by a virus, but this is not yet confirmed or generally accepted. However, it has been shown in some cases to be highly contagious from one individual shellfish to another.

The disease occurs throughout the world in a variety of bivalve molluscs and appears to cause significant mortality in certain farmed populations of shellfish.

Geographic Range and Species Infected

The disease affects many species throughout the world. Like many other shellfish diseases, it is probably more widely distributed than is now known. The following species and locations have been identified: *Adula californica,* Pacific coast of North America; *Artica islandica* (mahogany quahog), Rhode Island Sound, Atlantic coast of North America; *Cerastoderma edule* (common cockle), Cork Harbour, Ireland; *Saccostrea commercialis* (Australian rock oyster), Australia; *Crassostrea gigas* (Japanese or Pacific oyster), Matsushima Bay, Japan; *Crassostrea rhizophorae,* Brazil; *Crassostrea virginica* (Eastern or American oyster), Atlantic coast of North America, discontinuously from the Chesapeake Bay to Long Island Sound and sites on the Gulf coast; *Macoma calcarea,* Baffin Island, Canada; *Macoma nasuta* and *M. irus,* Yaquina Bay, Oregon; *Mya arenaria* (soft shell clam), Atlantic coast of North America, discontinuously from Chesapeake Bay to England; *Mya truncata,* Baffin Island, Canada; *Mytilus edulis* (bay or blue mussel), Pacific coast of North America, discontinuously from Yaquina Bay, Oregon, to sites in British Columbia, North Wales, Denmark, Finland, and southern coast of England; *Ostrea chilensis* (Chilean oyster), Chiloe, Chile; *Ostrea lurida* (Olympia oyster), Yaquina Bay, Oregon; and *Ostrea edulis* (European flat oyster), Mali-Ston area of Yugoslavia near Dubrovnic, Ria de Noya in Galicia, Spain, and the Brittany region of France.

Mortality Rate, Environmental Factors, and Seasonality

Some cultured populations appear to be 100% infected if individual animals are monitored over several months. Mortality rates due to the disease are reported to approach 100% over an annual period in some species. In other cases, in cultured populations, annual mortality rates of 30%-50% are typical.

Specific environmental factors that induce or enhance the disease are not known. Although much research has been conducted to determine whether various types of pollution contribute to the disease, no single factor which has these effects has been identified. Hemic neoplasia appears to be highly infectious, and dense populations of farmed shellfish maintain high disease levels because of ease of transmission from one animal to another. In all species in which seasonality has been investigated, the disease is reported to be most prevalent (judging by percentage of infected individual shellfish) during fall and winter months, typically from October through March. The prevalence drops in the spring and summer, possibly because heavily infected individuals die in the winter and the disease does not start another cycle of infection until autumn.

Diagnosis

Diagnosis is based on microscopic examination of blood or histological examination of tissues by a qualified pathologist. Common signs of the disease for all affected species are not established, but the following are known to apply in some cases: failure to produce mature reproductive follicles; high levels of mortality which spread geographically; and tissues swollen from the massive proliferation of abnormal blood cells in individuals with advanced cases of the disease.

Prevention and Management

Disease-free Areas

Every effort should be made to avoid introducing infected shellfish into areas that do not have the disease. Hemic neoplasia is known to be contagious from one animal to another within a given species. It is not known, however, if it can be transmitted from one species to another. It is possible, using techniques of pathological examination, to establish reasonable assurance of the presence or absence of the disease in given populations of shellfish. Field observations indicate that some populations within a given species may be more resistant than others.

Areas Known to Have the Disease

Eradication of the disease is not feasible since the disease can persist at low levels in natural populations of shellfish. General management methods based on available knowledge consist of keeping cultured population densities as low as practical and scheduling harvests so that market-sized shellfish are harvested before the typical period of increased disease in the fall and winter. Spat from wild spawn should not be allowed to collect on the shells of older infected bivalves.

It also appears that the severity of infection in individual shellfish and the percentage of individuals that are infected may increase as the animals get older. Thus, it may be desirable to harvest all individuals at as early an age as possible and to remove older shellfish from the population.

References

Alderman, D. J., P. Van Banning, and P. Colomer. 1977. Two European oyster *(Ostrea edulis)* mortalities associated with an abnormal hemocytic condition. *Aquaculture* 10:335-340.

Brousseau, D. J. 1987. Seasonal aspects of sarcomatous neoplasia in *Mya arenaria* (soft-shell clam) from Long Island Sound. *Journal of Invertebrate Pathology* 50:269-276.

Cooper, K. R., R. S. Brown, and P. W. Chang. 1982. Accuracy of blood cytological screening techniques for the diagnosis of a possible hematopoietic neoplasm in the bivalve mollusc, *Mya arenaria. Journal of Invertebrate Pathology* 39:281-289.

Cooper, K. R., R. S. Brown, and P. W. Chang. 1982. The course and mortality of a hematopoietic neoplasm in the soft-shell clam, *Mya arenaria. Journal of Invertebrate Pathology* 39:149-157.

Cosson-Mannevy, M. A., C. S. Wong, and W. J. Cretney. 1984. Putative neoplastic disorders in mussels *(Mytilus edulis)* from southern Vancouver Island waters, British Columbia. *Journal of Invertebrate Pathology* 44:151-160.

Elston, R. A., M. L. Kent, and A. S. Drum. 1988. Progression, lethality and remission of hemic neoplasia in the bay mussel, *Mytilus edulis. Diseases of Aquatic Organisms* 4:135-142.

Elston, R. A., M. L. Kent, and A. S. Drum. 1988. Transmission of hemic neoplasia in the bay mussel, *Mytilus edulis,* using whole cells and cell homogenate. *Developmental and Comparative Immunology* 12:719-727.

Farley, C. A. 1969. Sarcomatoid proliferative disease in a wild population of blue mussels *(Mytilus edulis). Journal of the National Cancer Institute* 43(2):509-516.

Farley, C. A. 1969. Probable neoplastic disease of the hematopoietic system in oysters, *Crassostrea virginica* and *Crassostrea gigas. National Cancer Institute Monographs* 31:541-555.

Farley, C. A. 1976. Proliferative disorders in bivalve mollusks. *Marine Fisheries Review* 38(10):30-33.

Farley, C. A., and A. K. Sparks. 1970. Proliferative diseases of hemocytes, endothelial cells, and connective tissue cells in mollusks. In *Comparative Leukemia Research 1969,* ed. R. M. Dutcher, Bibl. Haemat., No. 36 (Karger, Basel/Munich/Paris/New York), pp. 610-617.

Farley, C. A., S. V. Otto, and C. L. Reinisch. 1986. New occurrence of epizootic sarcoma in Chesapeake Bay soft shell clams, *Mya arenaria. Fishery Bulletin* (U.S.) 84(4):851-857.

Frierman, E. M., and J. D. Andrews. 1976. Occurrence of hematopoietic neoplasms in Virginia oysters *(Crassostrea virginica). Journal of the National Cancer Institute* 56(2):319-324.

Mix, M. C. 1975. Neoplastic disease of Yaquina Bay bivalve mollusks. In *Proceedings of the Thirteenth Annual Hanford Biology Symposium* 1:369-386.

Mix, M. C. 1983. Haemic neoplasms of bay mussels, *Mytilus edulis* L., from Oregon: Occurrence, prevalence, seasonality and histopathological progression. *Journal of Fish Diseases* 6:239-248.

Mix, M. C., and W. P. Breese. 1980. A cellular proliferative disorder in oysters *(Ostrea chilensis)* from Chiloe, Chile, South America. *Journal of Invertebrate Pathology* 36:123-124.

Oprandy, J. J., P. W. Chang, A. D. Pronovost, K. R. Cooper, R. S. Brown, and V. J. Yates. 1981. Isolation of a viral agent causing hematopoietic neoplasia in the soft-shell clam, *Mya arenaria. Journal of Invertebrate Pathology* 38:45-51.

Peters, E. C. 1988. Recent investigations on the disseminated sarcomas of marine bivalve molluscs. In *Disease Processes in Marine Bivalve Molluscs,* ed. W. F. Fisher, American Fisheries Society, Special Publication 18, pp. 74-92.

Twomey, E., and M. F. Mulcahy. 1984. A proliferative disorder of possible hemic origin in the common cockle, *Cerastoderma edule. Journal of Invertebrate Pathology* 44:109-111.

Vibriosis of Larval and Juvenile Molluscs

Vibriosis is an opportunistic disease of the larval stage of many, perhaps all, bivalve molluscs. It is also known to affect juvenile stages of the red abalone, *Haliotis rufescens.* The bacteria that cause the disease, members of the *Vibrio* group, occur in all marine waters where hatchery culture of bivalves is practiced. The disease is regarded by most as a "management disease," meaning that it can be controlled by proper hygienic procedures in the hatchery. In fact, the presence of the disease indicates that proper procedures are not being followed.

The disease has been reported since the beginnings of hatchery technology development. Many members of the *Vibrio* group of bacteria have not yet been specifically identified. It is likely that one or a few specific members of the group are the most important as larval bivalve mollusc pathogens.

Geographic Range and Species Infected

Vibriosis can occur in any marine hatchery situation since the causative bacteria are ubiquitous. Probably all species are subject to the disease, although some may be more susceptible than others. For example, the disease has been observed more often in American oysters, *Crassostrea virginica,* than in Pacific oysters, *Crassostrea gigas.* The disease in the red abalone is caused by *Vibrio alginolyticus,* one of the most common and widespread bacteria in the marine environment.

Mortality Rate, Environmental Factors, and Seasonality

In a well-managed hatchery there should not be any appreciable loss of larval bivalves to the disease under most circumstances. However, outbreaks can occur unexpectedly. In one well-documented case involving a production hatchery for the American oyster, a reduction of seed oyster production from 60 million (a good year) to 20 million oysters (a poor year) was attributed to vibriosis. The disease is associated with warm temperatures and typically is a problem only in the warmest months of the year.

Diagnosis

Vibriosis may be suspected when larvae grow slowly, batches of larval cultures fail, or larvae do not set. A confirmed diagnosis requires professional assistance and must be made by culturing the causative bacteria and examining the tissues of sick larvae. However, much can be done to detect vibrios in the hatchery and eliminate them from the system as discussed below.

Prevention and Management

Vibrios, like other pathogens, enter the hatchery or nursery by three principal routes: the seawater source, the brood stock, and algal food stocks. Since vibrios are ubiquitous, eradicating vibriosis is not possible and the disease is not an important consideration in the geographic transfer of larvae. However, good husbandry dictates that if animals are sick, from whatever cause, they should not be shipped, sold, or used for seed stock.

In a hatchery where vibriosis is suspected, personnel should become proficient at bacteriological sampling and culturing bacterial plates. A method for doing this is included elsewhere in the guide.

References

Brown, C. 1973. The effects of some selected bacteria on embryos and larvae of the American oyster, *Crassostrea virginica. Journal of Invertebrate Pathology* 21(3):215-223.

Brown, C. 1981. A study of two shellfish-pathogenic *Vibrio* strains isolated from a Long Island hatchery during a recent outbreak of disease. *Journal of Shellfish Research* 1:83-87.

Brown, C. 1983. Bacterial diseases in bivalve larval cultures and their control. In *Culture of Marine Invertebrates: Selected Readings,* ed. C. J. Berg, Jr. (Hutchinson Ross Publishing Co., Stroudsburg, Penn.), pp. 230-242.

Brown, C., and E. Losee. 1978. Observations on natural and induced epizootics of vibriosis in *Crassostrea virginica* larvae. *Journal of Invertebrate Pathology* 31:41-47.

Brown C., and G. Roland. 1984. Characterization of exotoxin produced by a shellfish-pathogenic *Vibrio* sp. *Journal of Fish Diseases* 7:117-126.

DiSalvo, L. H., J. Blecka, and R. Zebal. 1978. *Vibrio-anguillarum* and larval mortality in a California USA coastal shellfish hatchery. *Applied Environmental Microbiology* 35(1):219-221.

Elston, R. A. 1984. Prevention and management of infectious diseases in intensive mollusc husbandry. *Journal of the World Mariculture Society* 15:284-300.

Elston, R. A., and L. Leibovitz. 1980. Pathogenesis of experimental vibriosis in larval American oysters, *Crassostrea virginica*. *Canadian Journal of Fisheries and Aquatic Sciences* 37:964-978.

Elston, R. A., L. Leibovitz, D. Relya, and J. Zatila. 1981. Diagnosis of vibriosis in a commercial oyster hatchery epizootic: diagnostic tools and management features. *Aquaculture* 24:53-62.

Elston, R. A., L. Elliot, and R. R. Colwell. 1982. Conchiolin infection and surface coating *Vibrio*: shell fragility, growth depression and mortalities in cultured oysters and clams *Crassostrea virginica, Ostrea edulis* and *Mercenaria mercenaria*. *Journal of Fish Diseases* 5:265-284.

Guillard, R. R. L. 1959. Further evidence of the destruction of bivalve larvae by bacteria. *Biological Bulletin* 117:258-266.

Jeffries, V. E. 1982. Three *Vibrio* strains pathogenic to larvae of *Crassostrea gigas* and *Ostrea edulis*. *Aquaculture* 29:201-226.

Lodeiros, C., J. Bolinches, C. P. Dopazo, and A. E. Toranzo. 1987. Bacillary necrosis in hatcheries of *Ostrea edulis* in Spain. *Aquaculture* 65:15-29.

Tubiash, H. S., P. E. Chanley, and E. Leifson. 1965. Bacillary necrosis disease of larval and juvenile bivalve molluscs. I. Etiology and epizootiology. *Journal of Bacteriology* 90:1036-1044.

Tubiash, H. S., R. R. Colwell, and R. Sakazaki. 1970. Marine vibrios associated with bacillary necrosis, a disease of larval and juvenile bivalve mollusks. *Journal of Bacteriology* 103:272-273

Walne, P. R. 1958. The importance of bacteria in laboratory experiments on rearing the larvae of *Ostrea edulis* (L.). *Journal of the Marine Biological Association of the United Kingdom* 37:415-425.

Hinge Ligament Disease of Juvenile Bivalve Molluscs

In hinge ligament disease, the hinge or ligament that binds the two valves of a bivalve mollusc together is eroded or completely destroyed by bacteria. Known as "gliding" bacteria because of their distinctive motion, this specialized group of microorganisms has not been well studied. The gliding bacteria are known to have the ability to decompose many highly organized and complex biological structures made of protein, such as the hinge ligament of bivalve molluscs. Vast numbers of these bacteria are often found within the ligaments of juvenile clams, oysters, or scallops that sicken and die in nursery areas. Once the ligament is destroyed, the mollusc is unable to open its valves for feeding and respiration. Figure 9 shows in graphic form that these bacteria can cause the normally resilient ligament to soften and even liquefy at temperatures in the 5°C-10°C range and higher. It is also possible that the destruction of the ligament allows other bacteria to infect the tissues of the animals.

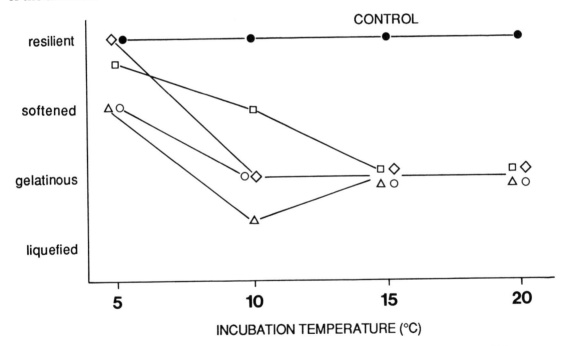

Figure 9. Effect of gliding bacteria on the hinge ligaments of the oyster. Four different gliding bacteria are compared with a fifth type of bacteria which is not capable of degrading the hinge ligament. The graph shows that gliding bacteria can cause softened, gelatinous, or even liquefied ligaments at temperatures down to about 10°C. The effects are less pronounced below 10°C. (Graph courtesy of C. Dungan)

This disease is the most important known disease of juvenile bivalve molluscs. It can infect any species. It has been reported from many locations where juvenile bivalves are intensively cultured. It is not known if the disease is important on oyster beds or among populations of wild oysters. The causative bacteria are probably common in all marine environments, and therefore the disease must be controlled by husbandry management techniques.

Geographic Range and Species Infected

Hinge ligament disease has been found in juvenile bivalves from both the east and the west coast of North America. It is likely that the disease occurs wherever bivalve molluscs are cultivated and potentially in any species. It has been found in the following species: Pacific oyster *(Crassostrea gigas),* Eastern oyster *(Crassostrea virginica),* European flat oyster *(Ostrea edulis),* hard clam *(Mercenaria mercenaria),* Manila clam *(Tapes philippinarum),* Pacific razor clam *(Siliqua patula),* and bay scallop *(Argopecten irradians).*

Mortality Rate, Environmental Factors, and Seasonality

In many cases, aquaculturists have reported the complete loss of large groups of clams and oysters from this disease. Usually the bivalves affected by the disease are from settlement size to 1 cm in shell height. The smaller animals are probably more susceptible.

No typical seasonal cycle of the disease has been determined. It can occur year-round, possibly because juvenile molluscs are usually grown in a controlled environment, often with heated water. Research on the disease has shown that the hinge ligaments are degraded at an increasing rate as the water temperature increases over the range from 5°C to 20°C and that the normally hard ligament, when infected with the gliding bacteria, can become jellylike at water temperatures as low as 10°C.

Diagnosis

There is no known way to make a certain diagnosis of hinge ligament disease without the microscopic examination of the ligament. However, in any large mortality of juvenile molluscs this disease should be suspect and samples submitted for a pathological evaluation.

Prevention and Management

Eradication of hinge ligament disease is not possible because the causative organism is common in marine environments. Thus, the disease must be prevented and limited by husbandry management techniques (see "Preventing and Managing Disease in the Hatchery").

The approaches tried have been geared toward the regular disinfection of the surface of juvenile molluscs. The most effective disinfectant has been sodium hypochlorite, otherwise known as common household bleach. Such a treatment can be applied only when the juveniles are in containers in a controlled environment, and the treatment is more practical for single bivalves than for oysters attached to cultch.

The concentration, frequency, and length of treatment may have to be adjusted to meet individual circumstances. A suggested starting point for the treatment is a three-minute dip in 25 parts per million sodium hypochlorite daily for five days. This concentration is made by diluting household bleach, usually labeled 5.25% sodium hypochlorite, with seawater by a factor of 2,100. This should be performed routinely if serious problems have been experienced from the disease. The continuing need for treatment will have to be determined for each operation.

Three antibiotics, penicillin, novobiocin, and tetracycline, are known to inhibit growth of some of the strains of causative bacteria. Penicillin is effective at restricting the growth of most, if not all, strains of the ligament-degrading bacteria, while novobiocin restricts growth of the least number of strains tested to date. Antibiotics are not recommended for routine use and should be applied only in a serious disease situation, with the dosage estimated in consultation with a pathologist.

References

Dungan, C. F. 1987. Pathological and microbiological study of bacterial erosion of the hinge ligament in cultured juvenile Pacific oysters, *Crassostrea gigas*. Master's thesis, University of Washington, Seattle.

Dungan, C. F., and R. A. Elston. 1988. Histopathological and ultrastructural characteristics of bacterial destruction of hinge ligaments in cultured juvenile Pacific oysters, *Crassostrea gigas*. *Aquaculture* 72:1-14.

Dungan, C. F., R. A. Elston, and M. Schiewe. 1989. Evidence for colonization and destruction of hinge ligaments of cultured juvenile Pacific oysters, *Crassostrea gigas,* by Cytophaga-like bacteria. *Applied and Environmental Microbiology* 55(5):1128-1135.

Elston, R. A. 1984. Prevention and management of infectious diseases in intensive mollusc husbandry. *Journal of the World Mariculture Society* 15:284-300.

Elston, R. A., L. Elliot, and R. R. Colwell. 1982. Conchiolin infection and surface coating *Vibrio*: shell fragility, growth depression and mortalities in cultured oysters and clams *Crassostrea virginica, Ostrea edulis* and *Mercenaria mercenaria*. *Journal of Fish Diseases* 5:265-284.

Ameboflagellate Disease of Larval Geoduck Clams

This disease of the larval geoduck clam *(Panope abrupta)* is caused by a parasitic protozoan known as an ameboflagellate and probably belonging to a group known as *Isonema*. It was recently discovered in hatchery-reared larval clams.

Geographic Range and Species Infected

The disease has been detected only in the one location where geoduck larvae are cultivated in Washington. Only geoduck larvae are known to be infected by this flagellate.

Mortality Rate, Environmental Factors, and Seasonality

Exact mortality rates of larvae due to the disease have not been determined, but hatchery personnel report losses to be "substantial." The mortalities have occurred throughout the time in which the larvae are cultured, from February through May.

Diagnosis

A preliminary diagnosis can be made microscopically by examining wet mounts of sick larvae for the characteristic protozoan (see Figure 10). The diagnosis should be confirmed by having a shellfish pathologist microscopically examine tissues for the presence of the parasites in the mantle and body cavity of the larvae.

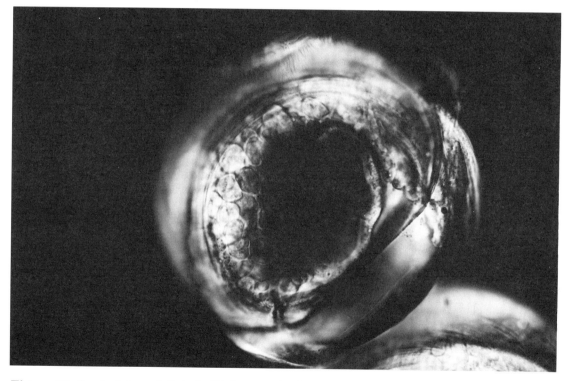

Figure 10. Geoduck larva infected with the parasite *Isonema*. The parasites are located between the valves (arrows) and can be identified with a low-power microscope.

Prevention and Management

No specific methods are known for the management of the disease. It is not known to infect juvenile or adult geoduck clams and does not infect oyster larvae grown in the same vicinity as the larval geoducks.

Reference

Kent, M. L., R. A. Elston, T. A. Nerad, and T. K. Sawyer. 1987. An *Isonema*-like flagellate (Protozoa: Mastigophora) infection in larval geoduck clams, *Panope abrupta. Journal of Invertebrate Pathology* 50:221-229.

Diseases of Abalone

The abalone group of molluscs, long cultured in Japan, are being increasingly farmed in North America. Several diseases have been reported, but relatively little is known about the diseases or health management of the abalone.

Vibriosis has been reported in the red abalone, *Haliotis rufescens,* as it has been in many intensively cultured molluscs. This disease is considered to be cosmopolitan, that is, caused by a common marine bacterium, and the effects of the disease can be controlled by good husbandry practices.

One of the most common bacteria in the marine environment, *Vibrio alginolyticus,* invades damaged epithelium or skin of the cultured abalone, then grows through the circulatory system of the animals, causing a fatal disease. Abalone stressed by high temperatures and supersaturated oxygen conditions are particularly subject to the disease. Nothing specific is reported for the management of this disease, but controlling temperature and oxygen levels as well as minimizing the number of vibrios in the system should reduce losses.

In Australia, a parasite known as *Perkinsus olseni,* similar to the one causing dermo (perkinsiosis) in oysters, is reported in the black-lipped abalone, *Haliotis ruber.* The disease is found in wild harvested abalone. It causes soft, cream-colored or yellow-to-brown pustules in the adductor muscle, in mantle tissue, and on body surfaces. Animals with these lesions are considered unacceptable for processing. Infection appears to depend, at least in part, upon temperature; abalone at 15°C had the lesions filled with dead parasites, while abalone at 20°C had active lesions containing live parasites.

Another parasite, *Labyrinthuloides haliotidis,* is reported to cause mortalities in hatchery-reared red and pinto abalone *(Haliotis rufescens* and *H. kamtschatkana)* in British Columbia. The parasite is considered to be a protozoan (single-celled animal). It was lethal to abalone under six months of age in an intensive culture facility. Parasites are found in the head and foot tissues of infected abalone. The parasites are reported to grow best at a temperature of 10°C (but not above 28°C) and a salinity of 30 parts per thousand in experimental studies.

Mortalities of cultured seed abalone of the three important species in Japan, *Nordotis discus, N. gigantea,* and *N. sieboldii,* have been attributed to bacterial pathogens. Mortalites were reported to be reduced by treatment with dihydro-streptomycin sulphate at 100 parts per million in the 32 days following development of the veliger larvae.

References

Bower, S. M. 1986. The life cycle and ultrastructure of a new species of thraustochytrid (Protozoa: Labyrinthomorpha) pathogenic to small abalone. Second International Colloquium on Pathology and Marine Aquaculture, September 7-11, 1986, Porto, Portugal, pp. 35-36.

Bower, S. M. 1987. *Labyrinthuloides haliotidis* n. sp. (Protozoa: Labyrinthomorpha), a pathogenic parasite of small juvenile abalone in a British Columbia mariculture facility. *Canadian Journal of Zoology* 65:1996-2007.

Bower, S. M. 1987. Pathogenicity and host specificity of *Labyrinthuloides haliotidis* (Protozoa: Labyrinthomorpha), a parasite of juvenile abalone. *Canadian Journal of Zoology* 65:2008-2012.

Bower, S. M. 1987. Artificial culture of *Labyrinthuloides haliotidis* (Protozoa: Labyrinthomorpha), a pathogenic parasite of abalone. *Canadian Journal of Zoology* 65:2013-2020.

Elston, R. A., and G. S. Lockwood. 1983. Pathogenesis of vibriosis in cultured juvenile red abalone, *Haliotis rufescens* Swainson. *Journal of Fish Diseases* 6:111-128.

Lester, R. J. G., and G. H. D. Davis. 1981. A new *Perkinsus* species (Apicomplexa, Perkinsea) from the abalone *Haliotis ruber. Journal of Invertebrate Pathology* 37:181-187.

Tanaka, Y. 1969. Studies on reducing mortality of larvae and juveniles in the course of the mass-production of seed abalone-I. Satisfactory result with streptomycin to reduce intensive mortality. *Bulletin of the Tokai Regional Fisheries Research Laboratory* 58:155-158.

LESS DOCUMENTED DISEASES

It should be emphasized that the absence of reported diseases for a particular species does not mean that the species is not subject to significant diseases. Molluscan species undoubtedly contract important diseases about which we know nothing. They also contract diseases that are well known but about which very little technical information is available as to their cause, prevention, and management. A number of diseases and parasites of molluscs are mentioned briefly in the technical literature but not in this guide, because too little is known about their relevance to mollusc culture or the effects on their host.

As more species of molluscs are farmed and as the production requirements for commonly cultured species become more rigorous, we will learn more about the importance of these diseases, their cause, management, and prevention. It is also important to note that diseases which are not important to one species in a given area can be important if they are introduced to a new host species or even to the same host species if it has not adapted to the disease organism.

Rickettsia and Chlamydia of Molluscs

Rickettsia and chlamydia are intracellular bacteria (that is, they live inside cells) that cause diseases in mammals, including man. Most bacteria, including those that can cause disease, do not actually reside inside living cells although they may live inside the host organism in various locations. Since there is no evidence that the similar organisms in molluscs cause diseases of mammals or man, they should be referred to as rickettsia-like or chlamydia-like.

This group comprises some of the most commonly observed microorganisms in the tissues of bivalve molluscs. They occur in healthy animals without causing any apparent detrimental effect. In several instances they have been blamed for massive mortalities of scallops, including the sea scallop (Placopecten magellanicus). This may eventually prove to be true, but further study on these diseases is required before we fully understand their significance. The microorganisms, essentially bacteria that are adapted to grow inside the cells of the host, are most commonly found in the epithelial tissues of the gills and digestive gland of the host bivalve mollusc.

These microorganisms occur in a variety of species of bivalve molluscs throughout the world. Chlamydia-like organisms have been reported in the bay scallop (Argopecten irradians), the Portuguese oyster (Crassostrea angulata), and the hard-shell clam (Mercenaria mercenaria).

Rickettsia-like organisms have been reported in the Pacific oyster (Crassostrea gigas), the Eastern oyster (Crassostrea virginica), Donax trunculus, the hard-shell clam (M. mercenaria), the soft-shell clam (Mya arenaria), the sea scallop (Placopecten magellanicus), the Pacific razor clam (Siliqua patula), the thin tellin (Tellina tenuis), the Manila clam

(Tapes philippinarum), the Japanese scallop (Patinopecten yessoensis), the European flat oyster (Ostrea edulis), and the Palourde clam (Ruditapes philippinarum).

References

Comps, M. 1982. Etude morphologique d'une infection rickettsienne de la palourde *Ruditapes philippinarum* Adam and Reeves. *Revue des Travaux de l'Institut des Pêches Maritimes* 46(3):141-145.

Comps, M. 1983. Infections rickettsiennes chez les mollusques bivalves des côtes françaises. *Rapports et Procès-Verbaux des Réunions Conseil International pour l'Exploration de la Mer* 182:134-136.

Comps, M., and R. Raimbault. 1978. Infection rickettsienne de la glande digestive de *Donax trunculus* Linné. *Science et Pêche, Bulletin de l'Institut des Pêches Maritimes* 281.

Comps, M., J. P. Deltreil, and C. Vago. 1979. Un microörganisme de type rickettsienne chez l'Huître portugaise *Crassostrea angulata* Lmk. *Comptes Rendus Académie des Sciences Paris* 289, Série D:169-171.

Elston, R. A. 1986. Occurrence of branchial rickettsiales-like infections in two bivalve molluscs, *Tapes japonica* and *Patinopecten yessoensis*, with comments on their significance. *Journal of Fish Diseases* 9:69-71.

Harshbarger, J. C., S. C. Chang, and S. V. Otto. 1977. Chlamydiae (with phages), mycoplasmas, and rickettsia in Chesapeake Bay bivalves. *Science* 196:666-668.

Meyers, T. R. 1979. Preliminary studies on a chlamydial agent in the digestive diverticular epithelium of hard clams *Mercenaria mercenaria* (L.) from Great South Bay, New York. *Journal of Fish Diseases* 2:179-189.

Morrison, C., and G. Shum. 1982. Chlamydia-like organisms in the digestive diverticula of the bay scallop, *Argopecten irradians* (Lmk). *Journal of Fish Diseases* 5:173-184.

Nuclear Inclusion X (NIX)

Nuclear inclusion X, or NIX, is a disease of the Pacific razor clam, *Siliqua patula*. It is caused by a highly specialized and very large type of rickettsia-like microorganism. It was first discovered on the Pacific coast in Washington in 1983 in association with a massive mortality of the razor clam. It infects the gill epithelial tissues and interferes with the respiratory processes of the clam. Virtually all clams in Washington are infected, as well as some populations in Oregon and British Columbia.

The disease persists at a low level in clams during the winter and spring. In some years the infection can greatly increase in intensity during the summer and fall, when mortalities associated with the disease usually occur.

Reference
Elston, R. A. 1986. An intranuclear pathogen [nuclear inclusion X (NIX)] associated with massive mortalities of the Pacific razor clam, *Siliqua patula*. *Journal of Invertebrate Pathology* 47:93-104.

Malpeque Bay Disease of the American Oyster

Malpeque Bay disease is a widely known but poorly understood disease that caused severe mortalities in American oysters *(Crassostrea virginica)* in Malpeque Bay in the Canadian maritime province of Prince Edward Island starting in 1915 and continuing through the 1930s. The geographical expansion of the disease, first observed a year after substantial plantings of seed oysters imported from the United States, is considered evidence for an infectious cause of the disease. More than 90% of original stocks were reported to have succumbed to the disease.

The oysters affected by the disease reportedly show visceral shrinkage, a translucent quality, reduced growth, and failure to spawn. The cause of Malpeque Bay disease has never been determined with certainty.

Reference
Needler, A. W. H., and R. R. Logie. 1947. Serious mortalities in Prince Edward Island oysters caused by a contagious disease. *Transactions of the Royal Society of Canada* 41(3,5):73-89.

Gill Parasite of the Japanese Scallop

The gill parasite, long recognized in Japan and described in the Japanese-language literature, was discovered in a group of scallops *(Patinopecten yessoensis)* proposed for introduction to North America. First thought to be a parasitic barnacle, this organism is now recognized as an unusual form of parasitic copepod, *Pectenophilus ornatus*. These raised yellow bodies on the surface of the gill can be as large as 8 mm in diameter (see Figure 11). In bottom culture in Japan, it was reported that up to 60 parasites can occur on an individual scallop. The number of parasites was greatly reduced in hanging cultures of scallops. This parasite, undoubtedly a burden to the scallop when it occurs in large numbers, is of more direct aesthetic significance, since a single parasite renders the scallop unacceptable as a whole animal product.

It was also reported to occur on another species of scallop in Japan, *Chlamys akazara*.

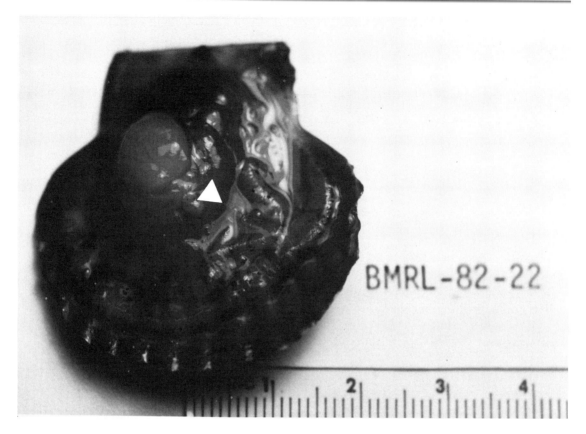

Figure 11. Japanese scallop infected with a single specimen of what is believed to be an unusual parasitic copepod, *Pectenophilus ornatus,* shown at the arrow. As many as 60 of these parasites can occur on the gills of one scallop, but a single parasite renders the scallop unmarketable as a whole animal product. (From Elston et al. 1985)

References

Gulka, G., P. W. Chang, and K. A. Marti. 1983. Prokaryotic infection associated with a mass mortality of the sea scallop *Placopecten magellanicus. Journal of Fish Diseases* 6:355-364.

Nagawawa, K., J. Bresciani, and J. Lützen. 1988. Morphology of *Pectenophilus ornatus,* new genus, new species, a copepod parasite of the Japanese scallop *Patinopecten yessoensis. Journal of Crustacean Biology* 8:81-42.

Miscellaneous Diseases

Mytilicola orientalis is a parasite found in the digestive tract of molluscs including several species of oysters, mussels, and slipper shells. It has been reported in Pacific oysters, *Crassostrea gigas,* in California, Oregon, and Washington. This parasite was introduced into France with imported Pacific oysters and is now present in the Arcachon region of France. *Mytilicola intestinalis* is a closely related species found in Europe.

The parasite can cause damage to the lining of the digestive tract where it attaches to the host. In Europe, it is reported that mortalities of mussels were caused by *M. intestinalis* when infestation reached levels of 5-10 parasites per mussel. Substantial mortalities associated with this parasite have not been reported in North American species of molluscs; however, infestations can lower the condition index of oysters.

Several viruses, in addition to those previously discussed, have been observed in the tissues of bivalve molluscs. These include a herpes-like virus in the American oyster, *Crassostrea virginica,* in the state of Maine. Preliminary studies suggested that virus was associated with mortalities at elevated seawater temperatures (28°C-30°C as compared with 18°C-20°C), but further studies would be required to prove that the virus caused the oyster deaths. Viruses similar to that causing velar virus disease in larval Pacific oysters, *Crassostrea gigas,* have been found in the blood cells and connective tissues of the adult Pacific oyster in France. It is not known whether these viruses cause disease or have any significant effect on the oyster.

"Australian winter disease" of the Sydney rock oyster, *Saccostrea commercialis,* is believed to be caused by a parasite known as *Mikrocytos roughleyi.* The disease was first reported in Australian oysters in 1926.

Parasites referred to as haplosporidans have been observed in a variety of molluscs. Two of the better known members of this group of parasites, *Haplosporidium nelsoni* (causative agent of MSX disease) and *Marteilia refringens* (causative agent of Aber disease), are discussed under separate headings.

Other members within this group of parasites appear to be important in causing diseases in many other molluscs, although they are less thoroughly documented. These less familiar diseases include an infection in gaper clams, *Tresus capax,* from Yaquina Bay, Oregon. The disease occurred in 43% of the clams, but only 20% had heavy infections in which clams were emaciated and sluggish and the mantle appeared watery and transparent. *Haplosporidium armoricana,* in Europe, is a parasite of the European flat oyster, *Ostrea edulis.* It appears to be well adjusted to that oyster species, occurring in fewer than 1% of oysters and causing little mortality. However, stocks of the Chilean oyster, *Ostrea chilensis,* introduced into France and exposed to *Haplosporidium armoricana* were infected to over 60% of the populations, with major mortalities.

Other haplosporidan parasites have been reported in the Pacific oyster, *Crassostrea gigas,* from Humboldt Bay, California, and the American oyster, *C. virginica,* from Tomales Bay, California, but whether or not these parasites cause any important disease in the oysters is unknown. Haplosporidan parasites are also found in several species of wood-boring bivalve molluscs.

Bucephalus haimeanus and *B. cuculus* are flatworm parasites of European flat oysters and American oysters. Early in the disease, white patches occur around the gonad area. Eventually, the entire reproductive and digestive tissue of the oyster is destroyed.

Nematopsis ostrearum and *N. prytherchi* are two gregarine parasites of *Crassostrea virginica*. Spores of the former are usually found in the mantle while those of the latter are found in the gill. Parasite-free oysters transferred into enzootic areas acquire the infection, but the infection is not lethal for the host and no sublethal effects have been documented.

References

Armstrong, D. A., and J. L. Armstrong. 1973. A haplosporidan infection in gaper clams, *Tresus capax* (Gould), from Yaquina Bay, Oregon. *Proceedings of the National Shellfisheries Association* 64:68-72.

Caty, X. 1969. Note préliminaire sur la présence de proliférations observées sur les huîtres atteintes de la maladie des branchies. *Revue des Travaux de l'Institut des Pêches Maritimes* 33(2):167-170.

Comps, M., J-R Bonami, and C. Vago. 1977. Pathologie des invertébrés: Infection virale associée des mortalités chez l'Huître *Crassostrea gigas* Thünberg. *Comptes Rendus Académie des Sciences Paris* 285, Série D:1139-1140.

Farley, C. A., W. G. Banfield, G. Kasnic, Jr., and W. S. Foster. 1972. Oyster herpes-type virus. *Science* 178:759-760.

Farley, C. A., P. H. Wolf, and R. A. Elston. 1988. A long-term study of "microcell" disease in oysters with a description of a new genus, *Mikrocytos* (g.n.), and two new species, *Mikrocytos mackini* (sp.n.) and *Mikrocytos roughleyi* (sp.n.). *Fishery Bulletin* (U.S.) 86(3):581-593.

Hillman, R. E. 1978. The occurrence of *Minchinia* sp. (Haplosporida, Haplosporidiidae) in species of the molluscan borer, *Teredo*, from Barnegat Bay, New Jersey. *Journal of Invertebrate Pathology* 31:265-266.

Katkansky, S. C., and R. W. Warner. 1970. The occurrence of a haplosporidan in Tomales Bay, California. *Journal of Invertebrate Pathology* 16:144.

Katkansky, S. C., and Warner, R. W. 1970. Sporulation of a haplosporidan in a Pacific oyster *(Crassostrea gigas)* in Humboldt Bay, California. *Journal of the Fisheries Research Board of Canada* 27:1320-1321.

Katkansky, S. C., A. K. Sparks, and K. K. Chew. 1967. Distribution and effects of the endoparasitic copepod, *Mytilicola orientalis,* on the Pacific oyster, *Crassostrea gigas,* on the Pacific coast. *Proceedings of the National Shellfisheries Association* 57:50-58.

Perkins, F. O., and P. H. Wolf. 1976. Fine structure of *Marteilia sydneyi,* haplosporidan pathogen of Australian oysters. *Journal of Parasitology* 62(4):528-538.

Sprague, V. 1949. Species of *Nematopsis* in *Ostrea virginica. Journal of Parasitology* 35:42.

Sprague, V., and Orr, P. E. Jr. 1955. *Nematopsis ostrearum* and *N. prytherchi* (Eugregarinina: Porosporidae) with special reference to the host parasite relations. *Journal of Parasitology* 41:89-104.

Taylor, R. T. 1966. *Haplosporidium tumefacientis* sp. n., the etiologic agent of a disease of the California sea mussel, *Mytilus californianus* Conrad. *Journal of Invertebrate Pathology* 8:109-121.

Needler, A. W. H., and R. R. Logie. 1947. Serious mortalities in Prince Edward Island oysters caused by a contagious disease. *Transactions of the Royal Society of Canada* 41(3,5):73-89.

Perkins, F. O., and P. H. Wolf. 1976. Fine structure of *Marteilia sydneyi,* haplosporidan pathogen of Australian oysters. *Journal of Parasitology* 62(4):528-538.

Sprague, V. 1949. Species of *Nematopsis* in *Ostrea virginica. Journal of Parasitology* 35:42.

Sprague, V. and Orr, P. E. Jr. 1955. *Nematopsis ostrearum* and *N. prytherchi* (Eugregarinina: Porosporidae) with special reference to the host parasite relations. *Journal of Parasitology* 41:89-104.

Taylor, R. T. 1966. *Haplosporidium tumefacientis* sp. n., the etiologic agent of a disease of the California sea mussel, *Mytilus californianus* Conrad. *Journal of Invertebrate Pathology* 8:109-121.

ANATOMY OF BIVALVE MOLLUSCS

An understanding of the basic anatomy of molluscs is a foundation for understanding how diseases can affect these animals. This section is intended as a reference to the major anatomical features of several bivalve molluscs. Additional information can be found in the following publications:

Barnes, R. D. 1987. *Invertebrate Zoology,* 5th ed. Saunders College Publishing, Philadelphia, Penn.

Elston, R. 1980. Functional anatomy, histology and ultrastructure of the soft tissues of the larval American oyster, *Crassostrea virginica. Proceedings of the National Shellfisheries Association* 70:65-93.

Galtsoff, P. S. 1964. *The American oyster.* Fishery Bulletin 48, U.S. Fish and Wildlife Service, 480 pp.

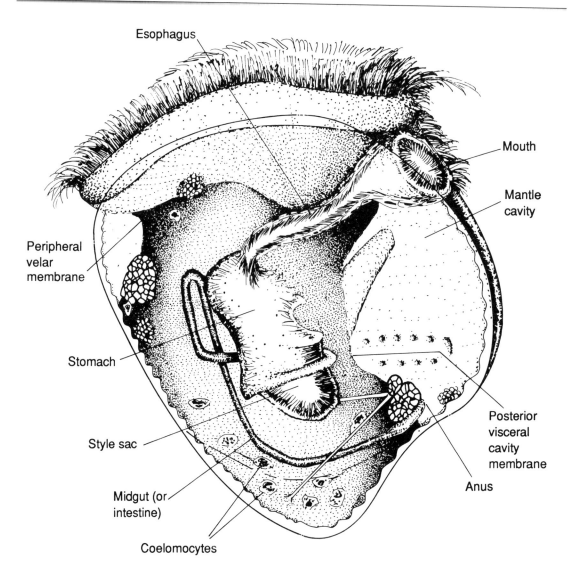

Esophagus

Mouth

Mantle cavity

Peripheral velar membrane

Stomach

Posterior visceral cavity membrane

Style sac

Anus

Midgut (or intestine)

Coelomocytes

Figure 12. Veliger larva of the oyster, measuring about 300 μm (equal to 0.3 mm, or about 12/100 inch) from the velum to the hinge. In this representation, the digestive gland and the retractor muscles are not shown in order to reveal the underlying structure of the organs. The anatomy of all veliger larvae of bivalve molluscs is very similar to that shown here. When extended from the valves, the velum is used for swimming, collecting food particles, and facilitating oxygen exchange. During settlement the animals undergo anatomical changes in which the viscera rotate within the shell, the adult gill develops, and other changes occur, including, in some species, the loss of the foot and one adductor muscle.

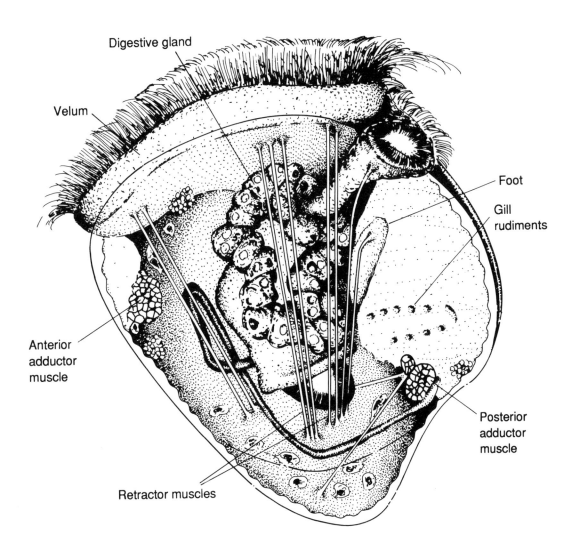

Figure 13. Complete veliger larva, similar to that shown in Figure 12 but with the digestive gland and retractor muscles shown in place.

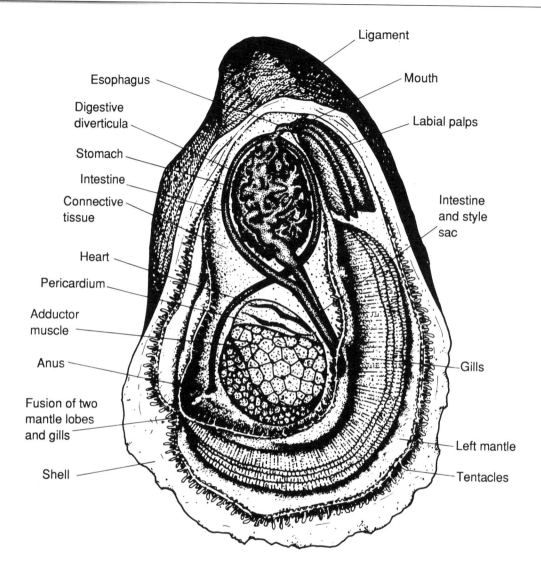

Figure 14. Adult American oyster *(Crassostrea virginica)* showing the anatomical reorganization mentioned in Figure 1. The mouth is now near the hinge region and, as in other oysters, the anterior adductor muscle has degenerated, leaving only the single larger posterior adductor muscle. After P. S. Galtsoff, *The American oyster,* Fishery Bulletin (U.S.) 64.

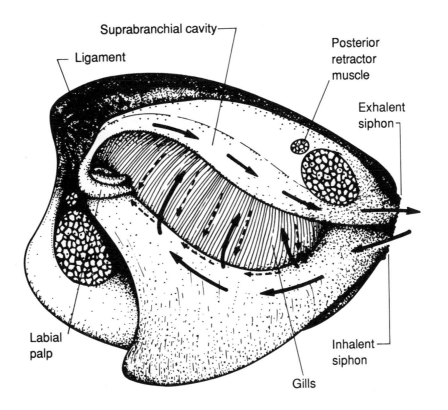

Figure 15. Hard-shell clam *(Mercenaria mercenaria)* with arrows showing the direction of movement of food particles and seawater over the gills. Redrawn from R. D. Barnes, *Invertebrate Zoology,* 5th ed. (Saunders College Publishing, Philadelphia, Penn., 1987).

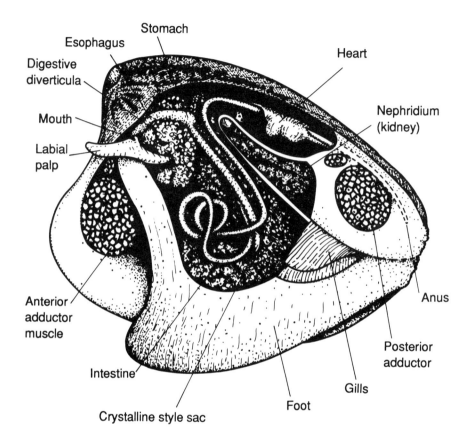

Figure 16. Hard-shell clam showing anatomical features. Redrawn from R. D. Barnes, *Invertebrate Zoology,* 5th ed. (Saunders College Publishing, Philadelphia, Penn., 1987).

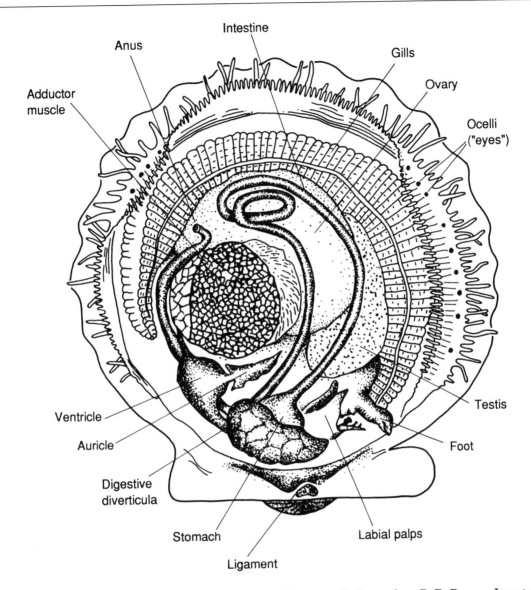

Figure 17. Sea scallop *(Pecten)* showing anatomical features. Redrawn from R. D. Barnes, *Invertebrate Zoology,* 5th ed. (Saunders College Publishing, Philadelphia, Penn., 1987).

PREVENTING AND MANAGING DISEASE IN THE HATCHERY

Diseases in the hatchery are caused either by *opportunistic* pathogens or by *host-specific* pathogens. These two kinds of diseases require different management approaches.

The diseases one finds most often in bivalve hatcheries are caused by bacteria that live in the marine environment whether or not bivalve larvae are present but that opportunistically take advantage of the high densities of larvae in the hatchery. Because of this, opportunistic diseases are often called *management* diseases, the commonest of which is vibriosis.

A host-specific pathogen, such as the virus that causes OVVD (oyster velar virus disease), can exist only in a specific bivalve species. Host-specific diseases are usually transported into the hatchery in live larvae or adult bivalves.

Based on the knowledge we have today, we generally consider most bacterial diseases of bivalve larvae to be management diseases. Viruses are host-specific pathogens, as apparently are at least some parasitic diseases (for example, the ameboflagellate disease of geoduck larvae). Having made this distinction, however, it should be noted that certain bacterial and fungal pathogens of larvae may turn out to be rather host specific, so that in reality there is not always a clear distinction between opportunistic pathogens and host-specific pathogens.

The reader may have noticed that most of the notable bivalve diseases are caused by parasites rather than by bacteria, fungi, or viruses. The bivalve parasites discussed in this book have long been recognized as important disease-causing agents in coastal environments. To date, however, they have not been recognized as direct causes of mortality in shellfish hatcheries. This may be a result of the fact that hatcheries are a fairly recent phenomenon; identifying parasites as a hatchery problem may be only a matter of time. It is also possible that some of the parasitic diseases affect or kill only adult bivalves rather than the larvae or juveniles found in a hatchery. Nonetheless, parasites, should they be recognized in a hatchery, can be managed by the principles outlined below for host-specific pathogens or opportunistic pathogens, as appropriate for the specific problem.

Opportunistic Diseases

Opportunistic diseases are caused by bacteria or other microorganisms that exist in the marine environment. Poor management practices can allow them to gain entrance at any number of sites in a hatchery, and good management can control or eliminate them. The overall approach to managing these diseases is as follows.

1. Maintain pathogen-free algal stocks and expanded cultures.
2. Maintain absence or low levels of vibrios and other disease-causing microorganisms in the system (water column and surfaces) by proper water filtration, hygiene of system

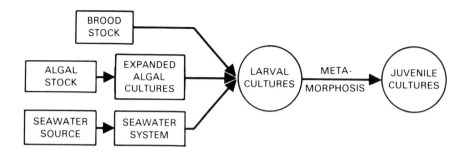

Figure 18. Intensive mollusc husbandry system showing how the components of larval culture combine to form the larval and juvenile cultures. Any component can become contaminated by bacteria or other microbes; thus, each component should be examined separately when trying to discover the origin of a problem in a hatchery. (From Elston 1984)

surfaces, and frequent water changes.

3. Isolate infected stocks and associated equipment at first sign of disease.

4. Discard infected stocks and sterilize equipment.

5. Identify source of contaminants, and modify and clean system. (See "Bacteriological Sampling.")

Host-specific Diseases

For host-specific pathogens (viruses and parasites), the following management practices should be followed:

1. Separate and isolate brood stock before, during, and after spawning until it has been determined that the brood stock or offspring are disease free.

2. Separate suspect larval and juvenile cultures from other cultures.

3. Use separate handling equipment and procedures for each stock and sterilize equipment after each use.

4. Destroy infected cultures and source brood stock if the presence of disease is confirmed.

5. Sterilize effluent from any brood stock or larval cultures that may carry a disease.

Bacteriological Sampling

Disease-causing bacteria can enter the hatchery system at any of its several components (shown in simplified form in Figure 18). To determine whether and to what extent bacteria are present, then, one must sample bacterial growth throughout the hatchery. Samples may be taken from water, container surfaces, or larvae.

Bacteriological sampling can be done at a several different levels of detail. It may be adequate simply to locate large numbers of likely problem-causing bacteria without actually counting bacteria. If, however, an actual count is needed, counting, while more detailed and time consuming, is possible in the production hatchery.

Once bacterial "hotspots" have been located and corrective measures taken (as outlined in the preceding section), the areas should be sampled again to determine the success of the measures.

The following procedures are directed toward the detection of vibrios, since vibriosis is the best-known bacterial disease, but they can be adapted to other bacteria. The steps outlined assume that the reader has some basic knowledge of bacteriological methods. Any hatchery worker who has successfully cultivated algal foods for feeding mollusc larvae will be able to perform these procedures.

Note: Dispose of used bacteriological plates carefully (by sterilization if this is practical) to avoid contaminating your system.

Supplies and Equipment

Whether you are simply *locating* bacteria or actually *counting* them, you will need the following supplies and equipment:
• bacteriological medium (see below).
• refrigerator for storing prepared medium and bacteriological plates.
• range top or gas burner for boiling medium; *or* autoclave for sterilizing medium.
• heatproof container in which to mix and boil medium.
• household bleach diluted 1:20 with tap water.
• standard bacteriological loops.
• plastic petri dishes (100 mm x 15 mm).
• weighing balance capable of weighing 1-gram quantities of medium.
• sterile cotton-tip swabs.
• gas or propane burner for flaming bacteriological loops.

If actual counts of bacteria are to be made, two methods are possible. For either method you will need the supplies and equipment listed above, as well as the following:
• sterile 10 mL capacity pyrex glass tubes with screw cap tops.
• bent glass rod (or "hockey stick") for spreading liquid samples evenly over the surface of an agar plate.
For Method 1, you will also need
• bacteriological "breed" loop.
For Method 2, you will need, instead,
• hand-held adjustable pipettor (with sterile disposable tips) capable of delivering 1 μL (microliter) increments up to a volume of 100 μL.

Medium

TCBS bacteriological medium (Difco Laboratories) is used to determine the presence of vibrios, many species of which appear yellow and cause the agar medium to turn yellow. TCBS medium is adequate for an initial examination where vibriosis is suspected in larval cultures.

Marine Agar (also from Difco Laboratories) is a medium which will grow a much wider variety of bacteria than TCBS. It may be used if other bacteriological pathogens are suspected or if "complete" counts of bacteria in the hatchery system are being made.

Other types of growth medium may be required for special purposes. For locating vibrio bacteria and roughly estimating their abundance, however, use the TCBS agar.

To prepare **Marine Agar,** follow label directions except that tap water can be substituted for distilled water to rehydrate the powdered medium. If the tap water is chlorinated, neutralize the chlorine or let the water stand at room temperature for 24 hours to allow the chlorine to escape into the atmosphere.

To sterilize the medium, use an autoclave at 15 pounds for 10 minutes. If an autoclave is not available, you can sterilize the medium by boiling it for 10 minutes on a range top. Medium containing agar boils over easily, so watch the mixture carefully once it reaches the boiling point.

When sterilization is complete, wipe down work surface areas with the 1:20 household bleach solution. Then carefully pour 15-20 mL of medium into each 100 x 15 mm petri dish using sterile technique. Allow the medium to remain at room temperature for 24 hours or longer to remove excess moisture. Package plates in sealed plastic bags and refrigerate until use.

To prepare **TCBS agar**, use a similar technique, except (1) follow label directions for boiling procedures and (2) use seawater diluted to about 10 parts per thousand or 1% sodium chloride instead of distilled water. Seawater can be diluted with tap water.

Taking the Samples

Bacterial samples may be collected from a number of places: the seawater source, any intermediate seawater holding tank, processed seawater as it is utilized, algal stocks, expanded algal cultures, and larval and spat tanks.

To collect samples simply to determine the presence and relative abundance of bacteria, run a swab over a small area if you are sampling a container surface; dip the bacteriological loop into the water if you are testing water. Transfer both samples onto the TCBS or Marine Agar plates by touching the swab or loop to the edge of the medium. Now spread the culture across the plate using a back and forth motion. Incubate the plates at an air temperature that is the same as the temperature of the culture water or, if this is not possible, at room temperature for 24-72 hours; and then compare cultures from different parts of the system and at different times.

For taking samples to be counted, use TCBS or Marine Agar, depending on the purpose. Collect the samples as follows.

Container Surface Samples

Using a cotton swab, collect samples from a defined square (1-2 cm on a side). This must be an area which has been removed from the water but not dried. Either drain tanks down slightly or use standard sheets of fiberglass, glass, or similar material which can be removed easily from tanks for surface sampling. Dilute sample by rinsing the swab thoroughly in a sterile tube containing 5.0 mL sterile seawater or 2.0% sterile saline solution.

(When sampling areas with very dense bacterial populations, a greater initial dilution or successive dilutions will be necessary.)

Plate the sample rinse from the swab onto the bacteriological medium (Marine Agar or TCBS) using a 0.01 mL breed loop (Method 1) or a pipettor (Method 2). Spread the bacteria with a glass rod. Incubate the plates at an air temperature that is the same as the temperature of the culture water, if possible, or at room temperature. Examine in 24-72 hours.

Water Samples

Water may be sampled by either Method 1 or Method 2. Using Method 1, sample container water with a breed loop and plate. Make a 1:500 dilution in sterile saline by transferring the breed loop sample to 5 mL of sterile saline, shaking, and sampling this with the breed loop.

If your conditions require further dilutions, they can be converted to numbers of bacteria per milliliter of water as follows.

The number of bacteria on the undiluted plate (if these are not too numerous to count) is multiplied by 100 (since the breed loop samples 1/100 mL) to give the number per milliliter. For the 1:500 dilution which has been sampled with the breed loop, multiply the number of bacteria on the plate by 50,000 (equivalent to multiplying by 100 and then by 500) to determine the number of bacteria per milliliter.

Method 2 is somewhat more flexible since the volume that is pipetted can be adjusted if an adjustable pipettor is used. Serial dilutions can be made as follows: Pipette 33 μL of the undiluted sample onto the agar surface and spread with the hockey stick. Next, dilute the sample 100-fold by transferring a 30 μL sample to a tube containing 3 mL of sterile saline. Next, shake this tube and transfer 33 mL from it to another agar plate and spread the sample. The number of colonies on the plate receiving the undiluted sample is multiplied by 33.3 to give the number of colonies per milliliter; the number of colonies on the 100-fold diluted plate is multiplied by 3,333 to obtain the number of colonies.

Larvae Samples

Samples of larvae can also be processed to give an indication of the bacteriological load of the larvae. This requires a glass tissue grinder ("Ten-Broeck" grinder) and sterile 1 mL and 5 mL pipettes.

Larvae are sampled by drawing larvae suspended in seawater up into a 1 mL pipette using an adjustable pipettor. Holding the pipette vertically, allow 0.1 mL of larvae to settle into the lowest portion of the pipette and dispense this amount into the sterile Ten-Broeck glass tissue grinder. This step must be performed very carefully in order to achieve consistent and useful counts from one sample to the next. Next, add 5.0 mL of sterile saline, and grind this suspension of larvae thoroughly. Make a series of quantitative bacteriological samples as described above with either a breed loop or an adjustable pipettor.

With careful technique and proper dilutions, you should obtain clearly separated, countable colonies. Many bacteria which spread or swarm (and are uncountable) on typical agar media (such as Marine Agar) will form discrete, countable colonies on TCBS.

SEEKING PROFESSIONAL ASSISTANCE

The diagnosis of many of the diseases of importance requires professional help. Unfortunately, very few individuals in the world today have training in shellfish pathology. They can be found in some government agencies, in some universities, and in some private organizations. Shellfish farmers will need to locate this help in their particular region.

A list of shellfish pathologists and pathology services appears at the end of this section. The list is not long. It is hoped that, as the recognition of shellfish health management grows, the number of professionals who can serve the industry will also increase.

In seeking the assistance of a shellfish pathologist or diagnostic professional, it is usually necessary to provide tissues for examination. These must be collected and delivered properly to be of use. One common mistake is failing to collect sick animals and tissues during a shellfish mortality. Often, a shellfish pathologist will be called to assist after the mortality has abated and no representative sick animals remain in the population. So, the first guideline is to enlist professional help *during* the actual problem, or at least to collect and chemically preserve tissues during this time.

The ideal way to deliver sick animal tissue for examination is fresh and within a few hours of collection, or to have a pathologist visit the mortality area to collect tissue and other samples. If this is not possible, representative sick animals should be delivered, on ice but unfrozen, to the examining pathologist within one day of collection. As a last resort, the tissues can be chemically preserved for testing, although fewer types of examinations can be done on preserved tissue than on fresh tissue.

Chemical Preservation of Tissues

There are many chemical solutions that can be used to preserve, or "fix," shellfish for pathological examination. Although a particular pathologist may have a preference, the following fixatives will be adequate. The simplest fixative should be used when there is not sufficient time to prepare the more complicated but preferred fixative.

Simplest fixative. The simplest fixative is formaldehyde, purchased as a 37%-40% solution and diluted at 1 part formaldehyde to 9 parts seawater.

Preferred fixative. The preferred fixative, referred to as "Davidson's" fixative, is prepared as follows. For 2 liters, combine and mix well:

600 mL 95% ethanol
400 mL 37%-40% formaldehyde
200 mL filtered seawater
600 mL tap water
200 mL glacial acetic acid

Whatever the fixative, shucked animals should be placed in the fixative with a volume of at least five times as much liquid as tissue mass.

In specific cases, the shellfish pathologist may require other types of tissue preparation, but this is an acceptable general method unless other specific instructions are given. Each container should be clearly labeled with the date and place of collection, the name of the species enclosed, and any other pertinent information.

Warning!

These chemicals are noxious. Use only with adequate ventilation. Do not let them come into contact with eyes or skin.

Also note that shells placed in Davidson's or any other acidic fixative release carbon dioxide and other gases. To prevent pressure from building up in the fixing vessels, do not seal the lids of the containers.

Shellfish Pathology Services

Practitioners are listed alphabetically by state. Do not send any material without contacting them in advance.

Dr. Theodore R. Meyers
Alaska Department of Fish and Game
FRED Division
P.O. Box 3-2000
Juneau, AK 99302
(907) 455-3597
Complete fish and shellfish disease diagnostic services for Alaskan facilities and for those out of state seeking certification of Crassostrea gigas *spat. No charge.*

Dr. Joe Sullivan
Alaska Department of Fish and Game
FRED Division
333 Raspberry Road
Anchorage, AK 99502
(907) 267-2249
Complete fish and shellfish disease diagnostic services for Alaskan facilities and for those out of state seeking certification of Crassostrea gigas *spat. No charge.*

Dr. R. P. Hedrick
Department of Medicine
School of Veterinary Medicine
University of California
Davis, CA 95616
(916) 752-3411
Oysters, abalone. Histology, $200 per 60 animals.

Dr. Carolyn Friedman
California Department of Fish and Game
Fish Disease Laboratory
2111 Nimbus Road
Rancho Cordova, CA 95670
(916) 355-0811
Bacteriology, parasitology. Preference given to government agencies, California-registered aquaculture and other aquaculture. No charge.

Dr. Walter Blogoslawski
NOAA, NMFS, NEFC
Milford Laboratory
212 Rogers Avenue
Milford, CT 06460
(203) 783-4235
Bacterial diseases of cultured oysters and clams. No charge; travel support required.

Dr. John C. Harshbarger
Dr. Esther C. Peters
Smithsonian Institution
Registry of Tumors in Lower Animals
NHB-W216A
Washington, DC 20560
(202) 357-2647
Neoplasms and related diseases. No charge.

Dr. John A. Couch
US EPA Environmental Research Laboratory
Gulf Breeze, FL 32561
(904) 932-5311
Toxicological pathology of molluscs, pathogenesis of parasitic infections (Perkinsus marinus, Haplosporidium), *neoplasia, etiologic agent diagnosis. No charge.*

Dr. James A. Brock
Aquaculture Development Program
335 Merchant Street, Rm 359
Honolulu, HI 96813
(808) 845-9561
General diagnostics for cold-blooded aquatic species. Supported by state of Hawaii.

Dr. Thomas C. Cheng
Marine Biomedical Research
Medical University of South Carolina
P.O. Box 12559 (Fort Johnson)
Charleston, SC 29412
(803) 795-7491 (or 7490)
Bacterial, protozoan, helminth, and arthropodan diseases; biochemical indicators of disease; large-scale surveys and epizootiological studies; consultation on preventive measures. Diagnostics, $250 per diagnosis.

Dr. Robert E. Hillman
Battelle Ocean Sciences
397 Washington Street
Duxbury, MA 12332
(617) 934-0571
Examination of shellfish stocks for evidence of parasites and pathogens. Sample of 50 individuals, $535.

Dr. Robin M. Overstreet
Dr. William E. Hawkins
Dr. Jeffrey M. Lotz
Gulf Coast Research Laboratory
P.O. Box 7000
Ocean Springs, MS 39564
(601) 875-2244
Molluscan and crustacean disease. No charge.

Dr. Robert E. Olsen
Oregon State University
Hatfield Marine Science Center
Newport, OR 97365
(503) 867-3011
Parasitology. Charge depends on service.

Dr. S. K. Johnson
Extension Fish Disease Diagnostic Laboratory
Department Wildlife and Fisheries Sciences
Nagle Hall, Texas A & M University
College Station, TX 77843
(409) 845-7471
General aquatic animal health and diagnostics; water quality management. Charge depends on service, usually under $25.

Dr. Sammy M. Ray
Ray Biological Consulting Co.
7213 Yucca Drive
Galveston, TX 77551
(409) 744-2761
Fluid thioglycolate culture analysis for Perkinsus marinus. *$10 per oyster.*

Dr. Eugene M. Burreson
Virginia Institute of Marine Science
Gloucester Point, VA 23062
(804) 642-7340
Protozoan parasites of oysters. No charge for Virginia residents.

Dr. Ralph Elston
Battelle Marine Sciences Laboratory
439 W. Sequim Bay Road
Sequim, WA 98382
(206) 683-4151
Complete fish and shellfish disease diagnostics and certification. Charges range from $200 to $1000, depending on service.

Dr. R. J. G. Lester
Department of Parasitology
University of Queensland
Brisbane, Australia 4067
(07) 377-3305
Protozoan and metazoan diseases of molluscs. No charge at present.

Dr. Susan M. Bower
Department of Fisheries and Oceans
Biological Sciences Branch
Pacific Biological Station
Nanaimo, BC, Canada V9R 5K6
(604) 756-7077
Parasites of abalone, scallops, oysters and clams on west coast of Canada; mussels, hemocytic neoplasia. No charge, but limited service as time and priority permit.

Dr. G. R. Johnson
University of Prince Edward Island
Atlantic Veterinary College
Diagnostic Services
550 University Avenue
Charlottetown, PEI, Canada C1A 4P3
(902)566-0864
Gross postmortem and histopathology, bacteriology for diagnostics and for depuration, marine toxin analysis (domoic acid), algal identification. Fee charged on a per test basis.

Dr. Takuo Sano
Laboratory of Aquatic Pathology
Department of Aquatic Biosciences
4-5-7, Konan, Minato-ku
Tokyo 108, Japan
(03) 471-1251
Aquatic pathology and virology. Charge not available.

J. F. McArdle
Fisheries Research Centre
Abbotstown, Castleknock, Dublin 15, U.K.
(01) 210111; Telex 31236 FRC EI ; Fax 205078
Diseases of farmed salmon and wild and farmed molluscs. No charge at present.

A. J. Figueras
Instituto Investigaciones Marinas (CSIC)
Eduardo Cabello 6
36208 Vigo, Spain
86 231930 / 86 292758; Fax 86 292762
Shellfish viruses, bacteria, and metazoan and protozoan parasites. Charge depends on sample size and frequency.

GLOSSARY

Antibody A protein produced in the body of a vertebrate animal in response to contact of the body with an antigen (an enzyme or toxin associated with a pathogen), serving to neutralize the antigen, thus creating immunity.

Bacteria One-celled microorganisms, larger than viruses, that occur in many forms. Some bacteria cause disease, but many are beneficial. (Singular = bacterium)

Cilia Hairlike outgrowths on the borders of some cells. (Singular = cilium)

Culture medium A substance, solid or liquid, used to grow microorganisms.

Electron microscopy Technique used to produce very highly magnified images of an object. The electron microscope focuses a beam of electrons through a magnetic field to produce a high-resolution image of an object on a fluorescent screen or photographic plate.

Enzootic Denoting a disease of animals which is indigenous to a certain locality, used synonomously with endemic.

Epithelium Cellular tissue covering surfaces, forming glands and lining most cavities of the body.

Epizootic Denoting a disease attacking a large number of animals simultaneously, or the prevalence of a disease; similar to an epidemic among humans.

Etiology The study of the causes of disease.

Flagellum A whiplike organ of locomotion in certain bacteria, protozoans, etc. (Plural = flagella)

Histology The branch of biology concerned with the microscopic study of the structure of tissues.

Host The organism in which or on which another organism grows and derives nourishment. (See **parasite** and **opportunistic**)

Immune system The complex of structures and functions of an organism that make it resistant to disease.

Indigenous Native; natural to area where found.

Infectious Designating a disease that can be communicated by contact with a disease-producing organism such as a virus or bacterium.

Lesion A pathologic change in the tissue.

Mantle The outer layer of tissue which enfolds all of the inner organs of a bivalve; the pallium.

Necrotic Denoting dead tissue.

Neoplasia The pathologic process resulting in the formation and growth of a neoplasm (a tumor, possibly malignant).

Opportunistic Denoting an organism capable of causing disease only in a host whose resistance is lowered, for example, by another disease.

Organism Any living individual taken as a whole.

Pallial surface The interior surface of the shell against which the mantle lies.

Parasite An organism that lives on or in another and draws its nourishment from it. The organism which is parasitized is termed the *host*.

Pathogen Something causing a disease, e.g., a virus, bacterium, etc.

Pathology Science concerned with the study of disease including the nature and cause as well as the structural and functional changes resulting from the disease process.

Prevalence The number of existing cases of a disease in a given population at a specific time.

Salinity The degree of saltiness of a substance. The salinity of oceanic seawater is about 32 parts per thousand.

Serological Relating to the branch of science concerned with serum, especially with specific immune serum. Serum is the fluid portion of vertebrate animal blood obtained after removing the fibrin clot and blood cells.

Sterile technique Method which is necessary to cultivate bacteria, viruses, and other microorganisms without extraneous contamination. Requires the use of sterilized instruments and culture containers, a clean isolated work place, and a gas flame for sterilization of instruments used in the procedures.

Ultrastructure Detailed microscopic structure or particles seen with the electron microscope.

Veliger The free-swimming stage of a bivalve larva.

Velum The veillike membrane that projects between the valves of a larval bivalve at the veliger stage. It bears cilia used for swimming, eating and respiration.

Virus A very small microorganism composed of an outer protein and a nucleic acid core. Viruses can grow and reproduce only within living cells.

Viscera The internal organs of the body.

Wet mount A preparation of living cells or tissue for microscopic examination, as opposed to cells or tissues that have been preserved and stained.

NATIONAL SEA GRANT DEPOSITORY
PELL LIBRARY BUILDING
URI, NARRAGANSETT BAY CAMPUS
NARRAGANSETT, R.I. 02882